APC Casebook:
Case Work Illustrations for
General Practice Candidates

Austen Imber

Routledge
Taylor & Francis Group

LONDON AND NEW YORK

First published 2004 by Estates Gazette

Published 2014 by Routledge
2 Park Square, Milton Park, Abingdon, Oxon OX14 4RN
711 Third Avenue, New York, NY 10017, USA

Routledge is an imprint of the Taylor & Francis Group, an informa business

British Library Cataloguing in Publication Data
A catalogue record for this book is available from the British Library

Library of Congress Cataloging-in-Publication Data
A catalog record for this book is available from the Library of Congress

ISBN: 978-0-7282-0567-3 (pbk)

Typeset by Amy Boyle, Rochester, Kent

Contents

Foreword

Understanding how the property industry works. That is the purpose of the Assessment of Professional Competence. *APC Casebook: Case Work Illustrations for General Practice Candidates* provides students with illuminating real-life examples of what the surveyor really needs to understand in order to progress.

These worked-through cases of lettings, rent reviews, lease renewals and the like are not only useful in themselves: they also help provide the student with a deeper understanding of the links between the subjects. It is these links – the effect of a letting on capital value for instance – that help the student appreciate the client's needs.

Estates Gazette hope you find this companion publication to *How to Pass the APC* a valuable volume – not only in helping with your APC, but also as a window into the way the sector really works.

Peter Bill
Editor *Estates Gazette*
June 2004

Preface

For graduates to be successful with the final assessment stage of the RICS Assessment of Professional Competence (APC), they require an applied understanding of the work undertaken in practice. This is gained by combining surveying experience with surrounding study, and being alert to the day-to-day issues arising in practice, and to the solutions that can be provided.

The APC is not about revising from text books, and an interview with three experienced surveyors is very different from the relative ease at which university exams can be passed.

Estates Gazette's first book on the APC, *How to Pass the APC: Essential Advice for General Practice Surveyors*, concentrated on the APC itself, and provided guidance on each stage of the process. A key point repeated on several occasions throughout the book was that an overall 65% pass rate (and approximately 50% first time pass rate) for general practice surveyors partly reflects the fact that many candidates sit the APC interview despite lacking the required experience and knowledge – including being unaware of some issues, and what a surveyor really needs to know.

APC Casebook: Case Work Illustrations for General Practice Candidates puts many areas of graduates' day-to-day case work into practice, and is an accompaniment to the whole of the structured training period – as well as being particularly helpful for final assessment.

Many of the illustrations are based on APC candidates' critical analysis, which is submitted as part of final assessment. The reports have been amended slightly in order to provide more detail in some areas, and to ensure that they do not simply represent template examples which can be mirrored by other candidates. This includes variations to RICS requirements, such as headings and word count (and it will therefore be important that candidates establish RICS requirements). The structure and essence of the critical analysis is, nevertheless, broadly maintained, and the reports show the importance of 'analysis', 'evaluation', etc, instead of the descriptive narrative that some candidates incorrectly present at final assessment. Some reports have however been amended slightly to be more factual, and in some cases this involves less extensive sections on critical appraisal/reflective analysis. Some elements of individual reports are also excluded because the points have been made in other reports. The first *Estates Gazette Books'* APC publication, *How to Pass the APC: Essential Advice for General Practice Surveyors*, provides guidance on how to prepare a good quality critical analysis, and also a further example of a critical analysis.

Within the various contributions of candidates' critical analysis, names and locations are disguised, and confidentiality sometimes further preserved through alteration of valuation figures and other aspects. Illustrations of interview questions are provided, although these are inevitably selective – and a wider range of questions could be received at the APC interview.

General practice graduates are typically involved in valuation, landlord and tenant, estate management, agency, rating and planning and development work. Valuation includes investment valuation, development valuation (which overlaps with development appraisal), rental valuation and an understanding of the Red Book. Some candidates may also undertake specialist valuations, such as in connection with licensed and leisure uses, or other properties whose value is dependent on business profitability. Landlord and tenant is concerned primarily with rent review and lease renewal, and also involves a thorough overall understanding of the landlord and tenant relationship, including ongoing property management issues. Estate management includes sub-lettings, assignments, surrenders, rent collection, dilapidations, dealing with other breaches of covenant, administering service charges, insurance, buildings/facilities management and other areas such as DDA compliance, managing asbestos, procuring works and ensuring compliance with health and safety legislation. Agency spans lettings on behalf of landlords and/or leasehold acquisitions on behalf of tenants of retail, industrial and office property, and any other specialist interests in which candidates may be involved. Sales and purchases could similarly cover a range of property types, and some candidates will be particularly involved in the investment and/or development markets. Planning and development includes commercial schemes and residential developments, and may also incorporate regeneration issues. Rating is another area in which candidates are often involved, and some candidates also gain experience in telecoms and insolvency work. Many of the above areas of work are reflected in the 17 chapters of *APC Casebook*.

One of many important aspects of APC candidates' training period is making the links between subjects, and understanding how the property industry works. Examples include the effect of lettings, rent review and lease renewal settlements on rental and capital value, and how lease terms can either help or hinder the ongoing estate management of a property.

Another important aspect of surveyors' work is understanding a particular client's objectives, and ensuring that property advice best fits their circumstances. There will be different objectives, for example, between a public sector property owner such as a local authority or utility, a corporate occupier such as a national retailer, and a private investment company or development company. Again, *APC Casebook* reflects such issues.

As with *How to Pass the APC: Essential Advice for General Practice Surveyors*, particular acknowledgment is due the following employers in view of their support for training initiatives for graduates other than their own employees.

Advantage West Midlands, the regional development agency for the West Midlands, and surveyors Honor Boyd, Kitt Walker and Karen Yeomans.

GVA Grimley, international property advisers, and training and development manager Scott Kind.

Advantage West Midlands and GVA Grimley also contribute to *Estates Gazette*'s 'Mainly for Students' series.

Input into *APC Casebook* has been provided by a number of former APC candidates. Honor Boyd and Kitt Walker are both Special Projects Managers with Advantage West Midlands. GVA Grimley's contributors are Piers Cartwright-Taylor, Rachael Jones, Sarah Kenney, Naomi Linnie, Kieran McLaughlin, Rupert

Mitchell, Roland Shaw and Kate Terriere. Jones Lang Lasalle's contributors are Guy Bransby and Rosie Young, and contributions are also made by Shrav Khera and Ruth Walsh. An APC event run by Birmingham Property Services is featured, and thanks here are due to BPS training manager Jacky Gutteridge and senior surveyor John Gaunt.

Thanks are due to Midlands Property Training Centre for its co-ordination of material, and ongoing help to general practice graduates as an independent facilitator of training support.

Also acknowledged is the support of Peter Bill, *Estates Gazette* editor, and Alison Richards, *Estates Gazette Books* Commissioning Editor, plus all those involved in the process – including Rebecca Chakraborty of *Estates Gazette Books*, and Adam Tinworth and Phil Brown regarding coverage of the APC in *Estates Gazette*'s 'Mainly for Students' series. Essential support in production has been provided by Audrey Andersson and Amy Boyle.

Austen Imber
May 2004

Chapter 1

Investment Valuation – Lending Purposes

This chapter examines an investment valuation undertaken for lending purposes. The surveyor has been instructed by the bank to report on the market value of the property, together with any factors affecting the suitability of the property for loan security. The report is contributed by Piers Cartwright-Taylor.

Introduction

I have chosen a case involving the valuation, for loan security purposes, of a freehold commercial property investment in Reading, Berkshire.

The client, a lending institution, was looking to use the subject property as security on a loan to their client, the borrower, who was in the process of acquiring the office building.

This particular valuation was chosen as there were a number of interesting lease terms and issues regarding the office market in Reading and the M4 corridor. It was also my first valuation carried out following the amendments to the RICS Appraisal and Valuation Standards, ie 'Red Book' in May 2003.

This report will discuss the valuation methodology and procedures followed in order to arrive at a value for the subject property. I will highlight some of the issues that I came across and the solutions to overcome these issues. Overall, the analysis will show how the valuation process has benefited my development as a general practice surveyor, and improved my knowledge of valuation techniques.

Valuation process

The following are the core elements that I identified in preparing this loan security valuation.

- Client's instruction.
- Property inspection.
- Market investigation.
- Quality of building/obsolescence.
- Lease analysis.
- Statutory enquiries.
- Methodology.
- Valuation and report.

Before the instruction could be accepted it was important to determine whether it was appropriate for the company to take on the instruction.

Understanding the client's requirement for loan security – key issue

The main objective is to provide the advice that is being requested. The bank's client intended to purchase the property, which was an income producing office building. We were required to provide the Market Value, as defined in Practice Statement 3.2 of the Red Book, and Market Value on the special assumption that the property is vacant.

When acting for a bank the question of suitability for loan security is vital since the bank will rely on the valuation in a future default scenario.

In compliance with my firm's ISO 9001 Quality Assurance procedures and Practice Statement 1.3 of the Red Book, it was important to establish whether any potential conflicts of interest existed. We found that my firm's rating department was, at the time, instructed to act on behalf of the tenant in respect of a rating appeal. This involvement was bought to the attention of my manager and also the rating department. On this occasion it was decided that no conflict arose. The rationale given was that the valuation was for our client, the bank, which was a sufficiently separate entity from the occupier.

Once agreeing that my firm could meet the client's requirements, the instruction was accepted in writing. Enclosed with the acceptance were duplicate copies of our conditions of engagement, drafted in line with Practice Statement 2 of the Red Book. This specified the purpose, basis and assumptions for the valuation, the name and qualifications of the valuer, the valuation date, time scale for reporting and the basis of fees.

Also, having regard to the RICS Rules of Conduct, I obtained the consent of the client and the lender to disclose the information outlined within this critical analysis. I have withheld their names in accordance with their instructions and for the purposes of confidentiality.

Property inspection

In accordance with Practice Statement 4, my supervisor and I inspected the property. The key factors noted are detailed below:

Location and situation

Reading is approximately 69 km (43 miles) west of central London in the Thames Valley, and benefits from good road communications. The property is situated on the northern side of Stevens Road, at its junction with London Street. Details on the location, situation, description and condition of the property are attached in Appendix IV (not reproduced here).

Description

The property comprised a detached office building arranged on basement, ground and six upper floors providing office accommodation on seven floors with ground floor and basement car parking. Photographs of the subject property are attached in Appendix V (not reproduced here).

Condition

The property showed few signs of disrepair. In any event, the lease is full repairing and the responsibility of repair rests with the tenant. This could be addressed by a schedule of dilapidations if need be.

Building obsolescence – key issue

Although the property was in a good state of repair, the age of the building was a matter of concern. At nearly 20 years old at the expiry of the reversionary lease, re-letting would become increasingly difficult with more modern space available on the market. This is reaffirmed by the level of the rent achieved in 1988, which has remained virtually unchanged.

Measurement

In conjunction with my supervisor, I measured the offices on a Net Internal Area (NIA) basis using a laser measuring device and in accordance with the RICS Code of Measuring Practice (5th ed). The building provided the following areas.

Floor	Use	m^2	sq ft
Basement	Car parking	693	7,460
Ground floor	Offices	71.9	774
Ground floor	Reception area	965.6	10,394
First floor	Offices	965.6	10,394
Second floor	Offices	965.6	10,394
Third floor	Offices	965.6	10,394
Fourth floor	Offices	965.6	10,394
Fifth floor	Offices	965.6	10,394
Sixth floor	Offices	965.6	10,394
Total		6558.5	70,598

The property had perimeter air conditioning units that were not of a continuous nature. In accordance with the Code of Measuring Practice, wall-to-wall measurements were taken.

Lease analysis and security of tenure – key issue

We were instructed to value the freehold interest, but were not provided with a Report on Title. However, the bank advised us that the property was freehold. For the purposes of the valuation, it was assumed that that there were no unduly restrictive covenants affecting Title, and which would have an adverse effect on value. The valuation report stated this assumption, and that it should be verified by the client's solicitors.

The property was subject to one occupational lease, producing a current rent of £1,145,000 pa. The lease had been granted in 1983 for a term of 25 years to a public company which had subsequently assigned the lease in 1995 to the Secretary of State. In 1996 a reversionary lease was granted to the Secretary of State for a period the period from 2008 to 2012. This meant that the question of privity of contract arose. Privity would apply to the original lease but not the reversionary lease since it is dated 1996, and is therefore covered by the provisions of the Landlord and Tenant (Covenants) Act 1995. A number of licences had been granted for alterations as the tenant was required to seek formal consent under the terms of the lease. A summary of key lease terms is attached in Appendix VI (not reproduced here).

Statutory enquiries

I undertook statutory enquiries to investigate whether there were any issues that may affect my opinion of value. Enquiries were made to Reading Borough Council and the Environment Agency in respect of:

- Planning.
- Highways.
- Rating.
- Environmental issues.

From my enquiries I was able to confirm that the property had consent for its existing use. I also found that neither the results of the planning nor highway searches revealed anything adverse to the client's interest. I visited Reading Borough Council who confirmed that the register identifying contaminated land uses has now been discontinued, since it has not found it necessary to take action concerning contaminated land, and accordingly they were unable to assist with our enquiries. I was unable to obtain a conclusive answer with regard to the contamination issues without obtaining a specific search. However, on the basis that the property is a town centre location without any nearby contaminative uses, the likelihood of contamination would be limited.

It was concluded that if an environmental audit was undertaken, and it was discovered that the property was contaminated, the valuation would be affected. This assumption was noted in the report to the client.

Investigation of the Reading office market – key issue

From my own knowledge of the local market, I was aware that there was a large over supply of offices in the Reading market and along the M4 corridor. My task was to obtain market evidence that was comparable to the subject property. This involved identifying recent lettings, together with freehold investment sales within the locality. First, I had to identify recent lettings in the locality which would help determine the property's Estimated Rental Value (ERV). Second, freehold investment sales would enable me to establish a yield that I could then apply in the valuation.

Collation of comparable evidence

I made inquiries with our agency department in respect of the Reading office market. I enquired as to whether they had recently let or acquired any properties on behalf of clients, or whether they were aware of any recent lettings or sales that may be suitable for comparable evidence. My initial thoughts were confirmed when I was told there was an oversupply of office space in the Reading market. I then searched for comparable evidence via external databases such as EGi and Focus.

Subsequently, I contacted agents who were active in the area, investigating any further evidence and the state of the Reading property market as a whole. These conversations gave me further information regarding the demand and supply for comparable property, including the duration of void periods and the level of letting incentives that were being offered.

Market conclusions

Rents in Reading town centre range from £15–£16 per sq ft for older second-hand accommodation to £24 per sq ft for modern recently completed developments. In the out of town markets, comprising Thames Valley Park and Green Park, and other smaller business parks, rents are typically around £26 per sq ft.

The office letting market is currently experiencing low levels of activity, although landlords are managing to maintain headline rents at high levels with the use of generous rent-free periods and other incentives. Due to long rent-free periods the headline rent being paid by a tenant can be misleading. When taking into account the overall rent payable over say, a five-year period the net effective market rent is much lower than the headline rent suggests. When analysing recent lettings I had regard to the level of rent-free periods granted, and calculated the net effective market rents. I enquired of my firm's rent review department as to the current practice when analysing net effective rentals. I was advised that a notional three-month fitting out period should be deducted from the rent-free period granted. The resultant period should be analysed over five years to give the net effective rental. With a large oversupply of offices, landlords were offering competitive rentals.

The majority of lettings over the past 18 months within Reading have been of small suites, whereas a letting in excess of 20,000 sq ft would be regarded as a large transaction. The subject property provides over 70,000 sq ft of accommodation, and it is probable that, in the current market, a discount on market rent for the quantity of accommodation in the subject property would be required.

(Notes were made of letting boards when inspecting the property, identifying both local and non local agents.)

ERV

Based on the evidence, I considered that the current rental achievable for the subject property as a headline rent would lie in the region of £1,410,000 pa (£20 per sq ft), from which a rent-free period of some 12 months would need to be granted.

Accounting also for three months' fitting out allowance, this equates to a net effective rent of some £16 per sq ft, or £1,129,568 pa.

On this analysis, the property is shown to be marginally overrented against a passing rent of £1,145,000 pa (£16.25 per sq ft). Also, the rent review effective from March 2003 was not actioned by the vendor. On balance, it was concluded that the offices were effectively rack rented. A summary of the key comparable rental transactions is included in Appendix VII (not reproduced here).

Investment market evidence and establishing a yield

The value of commercial properties depends upon factors including security of income, unexpired term of the lease, whether there are any break provisions for a tenant, the covenant strength of the tenant and the location and condition of the building.

The subject property is let to a government organisation, and this is considered to be the strongest covenant available. Yields for prime office investments let to good covenants with unexpired lease terms of over 10–15 years in Reading were found to be in the region of 7%–7.25%. However, the subject property is an older office building, let with an unexpired lease term of nine years. This would therefore warrant a discount on prime yields.

From our investigations, it became apparent that recent investment sales of properties in the Reading area were attracting yields ranging from 7.25% to 8.1%. Countrywide recently built office buildings which were let to the government for 20 years, and secured a yield of 6%–6.25%. In conclusion, the fact that the lease on the subject property had nine years to run, as well as taking into account the age, size and location of the building, discussions with my supervisor concluded that a yield of 8% was applicable. A summary of the key investment deals is attached in Appendix VIII (not reproduced here).

Valuation and methodology

The client bank required an independent valuer to advise them of the Market Value (MV) of the property, and the Market Value on the special assumption that the property is vacant.

The investment valuation was relatively straightforward, capitalising the rent of £1,145,000 into perpetuity at the above mentioned yield of 8%. Account also had to be taken of purchasers' costs, as yield analysis in the subject market and for the subject investment interests is on a net of costs basis (i.e. comparables are analysed on the basis of market rent divided by capital value/sale price plus purchasers' costs, and not simply market rent divided by capital value/sale price only).

The valuation was therefore as follows.

Rack rent	£1,145,000 YP perp 8% (12.5)	£14,312,500
	less purchasers' costs at 5.75%	£823,000
		£13,489,500
	say	**£13.5m**

As part of the standard reporting basis, and in drawing on the Kel valuation software package, the Nominal Equivalent Yield (NEY) was reported to the client. We also reported the Net Initial Yield (NIY). As the property was rack-rented (and as there was no need to build in any voids to the valuation), the NIY and NEY were the same, at 8.02%. We also reported the True Equivalent Yield (TEY), which was 8.44% – based on the rent being received quarterly in advance.

Quarterly in advance was in fact a possible valuation approach, but I opted for the traditional approach because of market practice. This was also necessary as it was important to work on with comparable evidence on a like for like basis.

The vacant possession valuation followed a similar format, but we increased the yield to 10% in view of the size of the building, the reduced covenant strength of more typical tenants, and the weakness in the office market. A void of 12 months was also built into the calculations. Although I was aware that evidence of vacant possession sales could be used, there was unfortunately no such evidence available.

Valuation and report

I drafted the report in accordance with Practice Statement 5 of the Red Book, following my firm's Quality Assurance procedure. A partner subsequently approved this. The report was presented in a format previously agreed with the client to suit their requirements.

Reflective analysis

A reflective analysis is provided below in relation to the key issues. Lessons learnt are then examined.

Understanding the client's requirement for loan security

I concluded the instruction within the agreed time scale, providing opinions of Market Value and Market Value on the special assumption that the property is vacant.

Quality of building/obsolescence

The issue of building obsolescence meant that the yield had to be adjusted to reflect the possible need for refurbishment and void periods that were likely to occur at lease expiry.

Security of income

The project identified the importance of analysing the key aspects of the lease and any terms that may have an effect on the value prior to undertaking the valuation. The most important item noted was that the reversionary lease that had been granted after the Landlord and Tenant (Covenants) Act 1995, and had implications regarding alienation. Where the government can assign, consideration of the alienation provisions is important in that privity of contract implies security of income under the original lease. In the event of the assignee defaulting under the terms of the reversionary lease, there would be no way in which the landlord could demand the rent from the original tenant, as the absence of an AGA meant that the government were no longer liable.

Investigation of the Reading office market

Investigation into the Reading property market was a key aspect of establishing the Estimated Rental Value. This was vital to provide sound valuation advice. Making thorough enquiries with local agents enabled me to truly understand the Reading property market.

Lessons learnt

I have learnt a number of lessons while undertaking this instruction that can aid me in the future and would summarise them as follows:

- The necessity to understand and meet the client's requirements is of great importance.
- During inspection of the property, comprehensive notes should be taken to include all factors that could affect the value of the property. They provide the information for preparation of the report.
- I learnt the importance of careful lease analysis. Inclusion of rent reviews and options to determine can have a significant effect on value. Additionally, the position regarding certain provisions, such as repairing obligations and options to determine, can have a material effect on value.
- Careful examination of the comparable transactions obtained is vital. Rents in Reading were of a broad range. Consequently it was required that adjustments be made to the evidence collated to derive my opinion of ERV and the appropriate yield.
- Teamwork is essential and extremely advantageous. The firm's agency department have extensive knowledge of the market and were able to advise as to the suitability of the rents and yields adopted within the valuation. Throughout the instruction, I was able to discuss matters with my supervisor and other colleagues within the department.

Appendices

The Appendices were: Appendix I – Location Map, Appendix II – Street Map, Appendix III – Ordnance Survey Map, Appendix IV – Location, Situation, Description and Condition of Property, Appendix V – Photographs, Appendix VI – Summary of Key Lease Terms, Appendix VII – Summary of Key Rental Transactions/Comparable Evidence, Appendix VIII – Summary of the Key Investment Deals/Comparable Evidence, Appendix IX – Kel Valuation (printout), and Appendix X – Red Book Definitions). None of the Appendices are reproduced here.

Illustration of issues/possible interview questions

Q Just to get us started, can you provide us with an overview of the performance of the Reading office market over the last 12 months?

Q What about rent-free periods – as you indicated, we often hear in the press that a certain level of rent has been agreed, but never know the extent of a rent-free period, if any, and how this might dilute the 'day-one', or 'effective', market rent. How precisely did you discount evidence where there was a rent-free period (an example can be provided if you wish)?

Q You indicated that a rent-free period for fit out would be treated differently. Why would this be the case, and would it always be the case?

Q 'M4 corridor' was a good term – what does this mean exactly, and what is the relevance of 'corridor'?

Q The Carsberg Report raised a number of issues which are now embodied in the new Red Book. What are these?

Q Can you please list illustrations of the application of the Red Book to this valuation case?

Q You mentioned the role of independent valuer. What alternative roles could there be?

Q Regarding the definition of 'Market Value', can you please define this in your own words (ie rather than a memorised version)?

Q What do you mean by the term 'reversionary lease'?

Q You mentioned that the reversionary lease was not caught by the privity of contract rules, and that the Landlord and Tenant (Covenants) Act 1995 applied because it was 'dated' in 1996. Is the criteria 'dated', or are there circumstances where the lease could be dated in 1996 and is still subject to privity?

Q You mentioned that the size of the property had an effect on its rent (and that quantum would apply). What sort of adjustments do you think would be made with lettings in the region of 10,000 sq ft?

Q Purchasers' costs were rightly deducted, but can you explain again why purchasers' costs are deducted in an investment valuation such as this?

Q You referred to the True Equivalent Yield and how this is the equivalent yield on a quarterly in advance basis, but why is this reported to the client, and why hasn't the quarterly in advance approach been embraced more widely in the market as valuation practice?

Q Regarding the vacant possession valuation, you applied an investment approach. How else could you approach this?

Q Looking at the valuation undertaken on the basis of vacant possession, what would be the advantages and disadvantages to investors if the property was vacant?

Q Can you give examples of where the vacant possession value might actually be higher than the investment value?

Q Did you investigate any alternative use potential?

Q What would be the costs of an investor holding a vacant investment property?

Q Had the rent passing been £900,000 and the market rent been £1.2m, with a rent review in three years' time and 13 years left on the lease, how might you have approached the valuation?

Chapter 2

Specialist/Profits Valuation – Healthcare

This chapter highlights some of the issues involved when properties are valued on their trading potential (as opposed to a conventional rate per sq ft and yield basis, for example). The specialist property concerned is a care home, and the report is contributed by Naomi Linnie.

Introduction

The case that I have selected is based on a valuation for loan security purposes of a converted care home in Cheshire.

While my specialist area is property marketing, I have more recently been involved in the valuation of healthcare properties, and have therefore chosen a valuation for my critical analysis. I selected this case because it encompasses the following:

- It reflects the work in which I have been involved.
- I was given a high level of responsibility and involvement which helped me to progress my skills in this area.
- The valuation reflects some of the key differences between healthcare and other property sectors.

The case selected is typical of a number of cases I have been involved with.

Objectives

We were instructed by our client (a bank) to carry out a valuation for loan security purposes, in order to enable the bank to provide funding to the purchaser for the acquisition.

Our client's principal objective was to establish the suitability of the purchase to support a loan.

The key issues include:

- Selecting the appropriate method of valuation for a trading care home.
- Ensuring compliance with RICS Appraisal and Valuation Standards, 5th ed (the Red Book).
- Inspection and statutory investigations – ensuring there were no issues that may effect the value of the property being acquired/valued.
- Analysis of the trading information provided, and collecting evidence from sales of other similar care homes.
- Analysis of comparable evidence and trading information from my firm's database of comparable transactions.
- Arriving at a valuation figure.

My personal objectives were to develop my experience and apply my professional skills in relation to the various competencies required for the APC, and to gain further valuation experience using the profits method.

Role and responsibility in the case

Under the supervision of a senior colleague in the department, I was given the responsibility for this case. It was important from the outset that I provided a professional service for our client, and that I was able to follow the procedural issues for Red Book valuation work.

Client confidentiality

In accordance with Byelaw 19 and Rule 9, Part III, of the RICS Rules of Conduct, the client gave consent that this case could be used as the subject of my critical analysis. Consent was also obtained from my firm and the purchaser.

Procedural requirements

A senior colleague received written instructions from the client to undertake the valuation, and in response I acknowledged the instructions and issued the standard terms of engagement. This included the terms and conditions and cost of the valuation, as well as generally complying with RICS requirements.

RICS Appraisal and Valuation Standards (Red Book)

All professional valuation work undertaken for loan security purposes is governed by the rules and regulations contained in the Practice Statements and Guidance Notes within the Red Book.

Quality Assurance

A file was opened in accordance with my firm's departmental Quality Assurance procedures, and the ISO 9001 Quality Assurance certification. A conflict of interest report was carried out using the internal instruction database system. No conflicts arose.

The property

The property is located in a town in Cheshire, and comprises a two-storey Victorian house. The building had been reconfigured, and extended to provide a 32-bed residential care home for the elderly. A purpose built single story extension was added in the early 1990s, and provides further bedroom accommodation. There is a lack of administrative office space within the main property, and a mobile temporary office on the grounds supplements this. Plans and photographs of the property can be found at Appendix A and B (not reproduced here).

Property inspection

I carried out an inspection of the property, making notes on the condition and number of bedrooms and en-suite facilities, bathrooms and lounge/dining areas. I also considered the general layout, and compliance with the requirements outlined by the Care Standards Act 2000 (CSA 2000), details of which can be found at Appendix D (not reproduced here).

As we were provided with a room disposition from the NCSC, a measured survey of the property was not required.

After the inspection I undertook an interview with the owner of the home to obtain all the relevant information needed to proceed with the valuation. This included:

- Floor plans.
- Recent certified trading accounts.
- A list of current residents and fees.
- A copy of the latest NCSC inspection report (unannounced).
- Copy of the registration certificate.
- Rates and hours worked by each grade of staff.
- Details of any dispute with employees, boundary disputes or issues on title.
- A copy of the home's statement of purpose.

The vendor also explained that his intention was to extend the home further. Plans had been drawn up, but planning permission had not been applied for.

When conducting the interview with the vendor it is important to establish why the home is on the market, as there may be factors which effect the reputation of the home. In this case the vendor was looking to retire.

Statutory enquiries

I made verbal enquiries to the local planning authority who confirmed the following:

- The property has permission for its current C2 use under the Town and Country Planning Act (Use Class Order) 1987.
- The property is Grade II Listed.
- The temporary planning consent for the temporary office had expired.

The planners also explained that, being a listed building, no further building work could take place that was directly connected to the original building. A stand-alone extension could be added, or an extension to the existing extension may be permitted. However, other issues, including parking provision, the surrounding amenity and the design and layout of the property would have to be considered prior to granting planning consent.

Although there is potential to develop further, without detailed planning enquiries and a planning consent it was not possible to assess accurately how many bedrooms could be added.

I made enquiries via the Homebuyers web site (www.homebuyers.co.uk) to ensure that there were no landfill sites within 250 m of the subject property.

I completed a search for all other care homes within a five-mile radius, which may be seen as competition. This was undertaken using a website which gives details of all care homes within the UK, and I cross-checked with the A–Z Care Home Guide. Although there were a large number of care homes within the vicinity, few homes offered the same category of care. Given the continued good occupancy rates at the home, I took the view that there was not an oversupply of places.

Valuation

Trading care homes are valued as trading entities, and the accepted valuation methodology is the profits method with appropriate cross-checks with comparable evidence and bench-marking. I referred to Guidance Note 1 of the RICS Appraisal and Valuation Standards regarding trade related valuation and goodwill.

The profits method involves an assessment of the maintainable profits that can be achieved by an average competent operator. This excludes any personal goodwill to the owner which would not be passed to the purchaser, but includes goodwill of the business.

The maintainable profits are subject to variables. The key variables are the fees, staff costs and non-staff costs, all of which are sensitive to quite modest changes.

The property

During my inspection, I made notes in relation to the physical condition and the spatial and environmental elements of the property. Of the 28 bedrooms, only five rooms have en-suite facilities. However, the overall quality and presentation of the property was to a reasonable standard, within a good residential area. This, coupled with the good reputation of the home, can account for the continued high occupancy rate, which is usually around 96%. One double room is used as a single room, giving an effective capacity of 31. The occupancy rate is therefore usually 100% of the effective capacity.

Fees

The average fee for the home is £320.87. 25% of the residents were privately funded, and the remainder were funded by Social Services. I contacted the local Social Services to enquire about the current fee rate, and the likelihood of an increase. They confirmed that the minimum fee rate would be increased by 2.5% in April 2004. I reflected this increase in my calculations of the Adjusted Net Profit, and estimated an average fee of £328.

Registration

The property is registered with the NCSC as a residential home for the elderly for 32 service users. I was provided with a copy of the latest NCSC report. Although the home was generally compliant with CSA 2000, various statutory requirements and recommendations were highlighted. Examples of these were that all radiators should be adequately protected (i.e. low surface temperature to prevent injury to the service user), and that windows on the first floor should have restrictors fitted.

Staff costs

I analysed staff costs using a staffing calculator which my firm has created using Microsoft Excel. Analysis of staffing showed that costs were approximately 42.6% of turnover. This is comparable to the market norm, as staffing costs for this type of care are usually in the region of 35–45%. Care home owners/managers are legally obliged to staff the premises at specified ratios as set out by the NCSC. I made adjustments to the staff costs to reflect the following points:

* The owner undertook the majority of the maintenance of the home, therefore the accounts did not reflect the cost of a handyman. A care home of this size would require approximately 20 hours of maintenance per week, at £5 per hour.
* The home did not have any laundry staff, and the night staff undertook laundry. The NCSC would request a member of staff to be dedicated to laundry if there was a change in ownership. I allowed for one member of laundry staff at 20 hours per week, at £4.50 per hour.
* The home allowed for three waking night staff, which is above standard requirements. I therefore reduced the number of night care assistants from three to two.

Adjusted Net Profit

Taking the above into consideration, I analysed the home's current EBITDA (Earnings Before Interest, Tax, Depreciation and Amortisation) and then calculated the projected ANP (Adjusted Net Profit) for the coming year. I allowed for inflation of non-staff costs at approximately 6%.

Period:	Actual year to 31/5/02	Projections, year commencing 1/10/03
Registration:	32	32
Effective capacity:	31	31
Assumed occupancy:	–	94%
Average weekly fee:	£320.87	£328
Revenue:	£481,054	£513,000
Staff costs:	£205,285	£230,000
Non-staff costs:	£86,655	£92,000
EBITDA/ANP:	£189,114	£191,000

Multiplying the maintainable profit stream (ANP) by the appropriate multiplier/years' purchase will give the value of the care home as a freehold operating business.

Rationale on selecting the multiplier

In reaching an appropriate multiplier, I looked at two care homes which my firm has recently sold. Both are in Cheshire, and provided the same category of care. Further details of these comparables can be found at Appendix C (see p19), and the multipliers for these homes are as follows:

	Adjusted Net Profit	Sale price	Multiplier
Property A	£198,780	£1,475,000	7.42
Property B	£197,219	£1,340,000	6.8

Following consultation with my supervisor, a multiplier ranging between 5 and 6 was applied to the sustainable ANP. The rationale for this is as follows:

- Property A is a 45-bed care home, which allows for better economies of scale.
- There is potential to create five further bedrooms.
- All bedrooms have en-suite facilities.
- Most of the service users are private clients.
- Property B is in a better location than the subject property, with close motorway access and a bigger catchment area.
- Generally the quality of the accommodation is to a higher standard.

Although the subject home trades with high occupancy levels, the home is tired in areas, and requires some redecoration. There are only five rooms with en-suite accommodation, and market trends show that most service users now expect to have an en-suite in every room. If a new purpose built care home were to be developed within close proximity, this may affect future occupancy rates.

I used the following table to display the effect of the multiplier:

ANP	Multiplier	Valuation
£191,000	5	£955,000
£191,000	5.5	£1,050,500
£191,000	6	£1,146,000

Through a discussion with my supervisor, I decided that the appropriate market value for this property is £1,100,000. This reflects the market, the attributes of the property and shortcomings of the subject home as outlined. This gives a multiplier of 5.7. The multiplier reflects the potential to extend the home (although limited) and hope value.

Alternative use valuation

When carrying out the valuation I also considered alternative use value. Due to the location and surrounding amenities, the most likely alternative use would be residential. I spoke to residential agents in the area, and undertook searches on the internet to establish the level of residential values within the vicinity of the subject property.

Although the town in which the property is situated is not generally a high value area, the particular location of the home is the most affluent part of the town. My research indicated an approximate alternative use value of £600,000 for residential use. This confirmed that the highest value for the property is as a care home. When reporting to the client, I stressed that in the event that the alternative use value should be important to the bank as part of its lending and risk assessment, then a valuer with the appropriate experience and local market knowledge should be instructed to provide a valuation accordingly.

Valuation rationale and loan security issues

In assessing the suitability of the property for loan security purposes I highlighted the following points to our client.

Strengths

- Strong demand for elderly care.
- Located within a high value residential area.

Weaknesses

- Listed building which has repairing obligations and limits the potential for extension.
- Converted building on two floors.
- Two shared rooms.
- Few bedrooms have en-suite facilities.
- Some bedrooms are less than 10 m^2 in area.

Opportunities

- Scope within grounds to extend the home, although planning approval may not be straightforward.

Threats

- A large number of care homes within close proximity to the subject property.

There has been an upward trend in care home values over the last 12 months, and while it is difficult to predict future trends, a continuation of this increase in values will, in part, be dependant on the degree to which state funded fee levels are increased in April 2004.

I concluded that, subject to the advice contained in the report, the property and business was suitable for loan security purposes at the value indicated, given realistic lending criteria.

Reflective analysis

I am of the opinion that all the objectives have been met. The project was carried out in a professional manner, and thus all aspects of the key issues were identified and fulfilled.

A number of issues have arisen that I have used to improve my knowledge and ability to undertake future instructions.

Although I had good comparable evidence available when carrying out the valuation, the absence of direct sale evidence is common in the healthcare sector. Owners tend to request confidentiality in order that the staff, referrers and NCSC are not aware that the property is on the market, for fear of disruption to the business. Many transactions are negotiated 'off-market'. It is therefore vital to examine closely the key trading variables prior to forming projections.

This highlights the importance of obtaining accurate information in the early stages. While undertaking this project, it became clear to me that it is very useful to get as much information as possible prior to the inspection and interview. This helps to conduct a more productive interview, and limits the necessity to go back to the vendor for further information.

In completing this case I have been able to improve my experience in many areas, including:

- Receipt and confirmation of client's instructions.
- Property inspection and interview with home owner.
- Collection of data which can impact on value.
- Valuation techniques.
- Understanding compliance with RICS requirements and my firm's quality assurance procedures.

I believe my personal objectives have been met. Key areas of professional development have been an increased confidence in client relations and an increased knowledge of valuation procedures and the factors effecting value. I have developed my knowledge of the sector and my confidence to communicate to clients, agents and other professionals.

In addition, I identified a number of issues while undertaking the instruction which will improve my ability to undertake similar instructions in the future. These include ensuring that:

- Copies of all relevant documentation are obtained and studied before inspection of premises.
- Where possible, time is allowed to complete all enquiries at the local authority, ensuring accuracy of information.
- Advice and assistance is sought, when required, from suitably experienced colleagues.

A number of general skills have been developed while undertaking this instruction which will also assist in my future work as a surveyor. These include:

- Report writing, letter writing and IT skills.
- Problem solving, evaluation and time management skills.

Since completion of this job, the Healthcare team have been instructed to carry out a portfolio valuation on behalf of the purchasers, who now own nine homes in total.

Appendices

The Appendices were: Appendix A – Location & Site Plan (1:1250), Appendix B – Photographs, Appendix C – Comparable Evidence, and Appendix D – National Minimum Standards. Appendices A, B and D are not reproduced here.

Appendix C – Comparable evidence

Property A, Cheshire

Description: 45-bed carehome. 30 rooms, 28 with en-suite. Potential to create five further rooms. Good location in prosperous catchment area. Average fee £370 per week – mainly private clients. Registered for 45.

Financial Details:

Income:	£471,237
Total staff costs:	£164,713
Net profit:	£97,132
EBITDA:	£178,444
ANP:	£198,000

Sale price: £1,475,000.

Property B, Cheshire

Description: 27 rooms, all en-suite. Located in prosperous village in Manchester commuter belt. Average fee £430 per week – mainly private clients. Registered for 30.

Financial details:

Income:	£471,856
Total staff costs:	£169,000
Net profit:	£151,398
EBITDA:	£175,618
ANP:	£197,000

Sale Price: £1,340,000

Illustration of issues/possible interview questions

Q You refer to legislation regarding care standards. I understand that a number of care homes have closed in recent years because of the cost of upgrade not being viable. What are the current market trends here?

Q Can you please summarise the key elements of the Red Book which might have particular application to a loan security valuation of a care home (as opposed to factors of general application to loan security valuation)?

Q The occupancy rates seem high at 94%. I do not doubt this for the property and its current operator, but is mid-90s the market trend in the area?

Q You mentioned that maintainable profits exclude any personal goodwill to the owner which would not be passed to the purchaser – was there anything special about the current owner's operation of the business which would not be available to the market?

Q What are the general trends with staff costs, and what sort of operations may involve particularly high staff costs?

Q Were there any principal defects noted on the property inspection – and even if not, what might typical defects be for properties of this age and construction?

Q Your valuation is really of an operational business, and you do not indicate a breakdown between, say, the property/bricks and mortar, fixtures/fittings/furniture/equipment etc. and any goodwill of the business. Why does this not need to be done?

Q What stamp duty is payable – the usual 4% assuming a property transaction, or something else?

Q Are there any implications regarding DDA with such properties?

Q What was your view on likelihood of planning permission being obtained for an extension, and remind us how this potential significantly affected your valuation?

Q You mentioned 'hope value'. Can please you explain that in more detail?

Q Comparable A has got 45 beds, but you state '30 rooms with 28 en-suite'. Does this mean that some residents share a bedroom, and how is this reflected in fee rates and any valuation adjustments? Also, the income looks low for the number of beds – why was this?

Q EBITDA – can you explain what exactly this is, and whether it differs from the usual 'net profit' figure in company accounts (ie what the deductions will be in order to more accurately establish how much money a new owner may make, rather than a basis which is helpful for valuation consistency)?

Q Regarding your EBITDA/ANP of approx. £190,000, does this include all of the proprietor's remuneration, or a salary included in the costs?

Chapter 3

Rating Appeal

This chapter outlines the process and issues arising in connection with a rating appeal on retail premises occupied by a bank. The report is contributed by Rosie Young.

Introduction

The aim of this critical analysis is to examine the business rates liability for a standard retail unit, and illustrate both the processes and influencing factors that have resulted in obtaining a reduction in the property's Rateable Value (RV). Furthermore, it will evaluate both the success of the project and examine the skills and lessons learnt.

This particular property was chosen because I have had a substantial involvement in most stages of the instruction, and also because I feel that it is representative of some of the common issues that impact upon the rates liability for retail premises.

This project examines the non-domestic rates liability of a 1960s retail unit located in Kingswinford, West Midlands (see Location Plan – Appendix I – not reproduced here) and illustrates how a reduction in the Rateable Value on the 2000 Rating List was achieved.

The subject property is currently used as a bank and is fully occupied by our client, a major high street bank. It comprises a ground floor banking hall with office accommodation to the rear and ancillary storage accommodation located on the ground and first floors (see Description – Appendix II – not reproduced here).

It is currently let on a 40-year full repairing and insuring lease from 1 January 1969 with five-yearly rent reviews (see Appendix III – Lease Summary – not reproduced here).

Procedures adopted

The reason for our instruction was to establish the correct RV for this property, which in so doing would hopefully minimise our client's business rates liability. It was necessary to establish whether the existing compiled list entry represented a fair assessment for the property and take any necessary action if not.

To establish if this was the case the following steps were undertaken:

- Researching the business rates history for the property and discovering whether the liability was affected by the government's transitional provisions.
- Inspection and measurement of property.
- Obtaining and understanding tenure information.

- Identifying the rateable hereditament and checking the extent of our client's liability.
- Valuation considerations.
- Client reporting.

Key issues

By completing these steps a number of key issues were highlighted in the process. It was apparent that there was a large disparity between the passing rent and the RV. I needed to identify why this was the case and establish the correct rental value for this property. Upon examination of the initial valuation provided by the Valuation Officer (VO), it also became apparent that the zoning approach that had been adopted was incorrect. Difficulties also arose in agreeing on the Zone A rate to be adopted, because the rent on this property suggested a considerably lower rate than many of the comparables used to establish the tone of the list.

Supervision

Although this project was allocated to me to undertake and complete, I was assigned a supervisor within the rating department who knew the property, and was aware of the pertinent issues. Regular meetings and discussions were held with my supervisor to ensure that I was addressing all these issues, and someone was available to answer my queries.

Furthermore, by following the firm's Quality Assurance system I was able to comprehensively detail my work in a transparent manner, thereby enabling my supervisor to review the work I had undertaken.

Permission has been sought and obtained from the client who I was acting for on this appeal. Both my firm and the client are aware that I am using this case for the purpose of my critical analysis, as part of the requirements for the Assessment of Professional Competence, and that the information contained will solely be used for this purpose.

Options and methodology

My involvement in this case began approximately two months prior to the Valuation Officer's 'target date for settlement'. A proposal had already been lodged against the compiled list entry figure of £19,500 rateable value before the initial deadline at the end of September 2000. This ensured that any alteration to the RV would have an effective date of 1 April 2000, thereby maximising the benefits of any rate savings obtained for our client.

The terms of engagement and fee basis had already been established, as well as a conflict of interest check. This highlighted that my firm had undertaken the latest rent review in 1999, so the file was obtained.

Rating history

My first action was to establish the history and rating background of the property. I discovered that the appeal against the 1995 RV of £12,300 had been withdrawn, and confirmed that the current RV was £19,500. By doing this, I was able to establish how much growth there had been between rating revaluations, as well as ensure that the RV to which our appeal related had not subsequently been altered by a Valuation Office Notice.

I was also able to check whether the liability was affected by the government's transitional arrangements. By using our 'in house' rating computer software, I established that the liability was subject to transitional relief, limiting the increase in charge from 1 April 2000. This was important because it helped me to establish how beneficial any reduction in RV would be for the client with regards to rate savings. The business rates liability for this property, before any appeal was made, can be seen in Appendix IV (not reproduced here) and shows the property in transition for the first three years.

Inspection and measurement

I liased directly with the branch manager and inspected the property on 26 February 2003.

The property was measured on a Net Internal Area (NIA) basis in accordance with the RICS Code of Measuring Practice (5th ed). This enabled me to calculate the floor areas and zone the property correctly. I used the NIA basis of measurement because this is the accepted method for retail premises. By adopting the same approach as the VO, it meant that comparisons could be made, enabling any factual differences to be identified and resolved at an early stage in the negotiation process.

While on site I noted the standard of accommodation and any additional rateable items. It was noted that there were two wall mounted air-conditioning units. I was also able to see if the property had any disabilities that may have had an impact upon the rental value.

Being on site provided me with the opportunity to discuss with the occupier whether there had been any alterations, extensions or demolitions to the property and generally extract as much information as possible.

An external inspection was undertaken to check that there was no car parking, and to establish the property's pitch in relation to the other retail units in Kingswinford. The frontage for the unit was reasonable, although there was a degree of masking to the Zone A area which was noted – as was the existence of one step up into the property. A goad plan was drawn and a note was made of any agents' boards in the area. These actions were undertaken for the purpose of obtaining comparable information.

Once the floor areas had been calculated, I could compare mine to those provided by the VO. It was clear that both sets of floor areas were similar, and that this would not be an issue in the negotiations. By establishing this early on in the negotiation process, I had given myself time to address any factual differences that may have arisen and if necessary re-inspect.

The floor areas on a NIA basis were as follows:

	Metric	*Imperial*
Ground – Zone A	62.74 m^2	675.36 sq ft
Ground – Zone B	79.64 m^2	857.24 sq ft
Ground – Ancillary	77.13 m^2	830.23 sq ft
Ground – Store (external)	4.18 m^2	44.99 sq ft
First – Ancillary	45.21 m^2	486.64 sq ft
Total	268.90 m^2	2,894.45 sq ft

Tenure information

A detailed summary of the lease is provided in Appendix III (not reproduced here).

The rent review file showed that there had been a nil increase at the last rent review, and the passing rent was £9,750 with effect from 1 January 1999. This suggested that there had been little, or no, rental growth in the area, or that the property had previously been overrented. A brief comparison of the rent and RV showed a greater difference than I would have expected, and I needed to establish why this was the case. From my knowledge of the local area, I knew that retail markets in many of the regional towns in the Black Country had been weakened by the opening of the Merry Hill Centre. This had a detrimental effect on rental levels, and could have accounted for the nil increase at review of this property. From a rating viewpoint however, I was able to argue that due to the downturn in the market, there had been no change between the lists and the RV should not have increased.

Identifying the hereditament

I decided that I needed to identify the demise that the rent related to, and compare this with the rateable hereditament. Upon examination of the original lease, I was able to determine that the rent related only to the front portion of the property, and that the unit had been considerably altered and extended by the tenant since the lease commenced.

For ease, I marked the original demise on a plan and calculated the area in m^2. The VO was aware of the passing rent however, there was disagreement regarding the area to which the rent related to, and this was important in the valuation of the property. After discussions with the VO, we were able to agree exactly what the rateable hereditament comprised and the area the passing rent related to.

Valuation considerations

In assessing the property the VO adopted a rate of £125 per m^2 ITZA. By examining other assessments in the area it was clear that this was considerably lower than most other units which I believed to be due to its off pitch location. Despite this I was still of the opinion that the adopted main rate was too high and did not represent an accurate rental value.

I felt that there were two key issues in establishing the correct level of value for this property. First to establish how strong the market was or how badly it had been affected by the Merry Hill Centre by examining and analysing comparable evidence. Second, I wanted to establish why there was such a difference between the RV and passing rent and ensure that the valuation approach for this individual property was correct.

An analysis of comparable evidence can be seen in Appendix V (not reproduced here). When I began my research into what had occurred in the area, it became obvious that the tone of the list had been well established. This was because the majority of appeals had already been dealt with on a previous listing, with only the A2 users remaining. The bank was situated next to a library (which was not a suitable comparable) and a local restaurant (which had not appealed their assessment). The nearest A2 user was a Lloyds TSB opposite, but again, the assessment had not been appealed. The HSBC bank also on the same street was valued at a rate of £250 m^2 ITZA, and this level had been supported on the neighbouring property at Valuation Tribunal.

My next step was to analyse the passing rent on the subject property. Based on my areas and relativities, I arrived at a Zone A figure of £107.81 per m^2. Adopting the VO's areas and relativities, the rent devalued at a rate of £93.75 per m^2 ITZA. This was one year after the Antecedent Valuation Date (AVD) of 1 April 1998, suggesting that some rental growth may have occurred. However, I still felt the main rate adopted was excessive, especially in light of the downturn of the retail market in recent years that had been experienced at this locality.

The VO was reluctant to accept the rent as the best evidence of value, because it suggested a considerably lower value than the existing tone of the list. After lengthy negotiations and careful examination of the comparable evidence, I was able to persuade the VO to adopt a rate of £110 per m^2 ITZA.

There were also a number of specific issues relating to this property that required discussion. Originally the VO had uplifted the office areas and applied a discounted office rate to these areas. I highlighted the fact that there was no separate access to these areas, and therefore did not believe they could be let as offices on the open market. Through negotiation, I was able to persuade the VO to adopt a traditional zoning approach, and zone through the ground floor. I also persuaded the VO not to uplift for A2 use by increasing the rate applied to the strong room, as smaller banks such as this no longer use these areas for anything other than storage.

I examined the frontage and disability issues possessed by the property. By looking at the guidelines laid down by the Inter Banking Rating Forum (IBRF), I placed the frontage as a Category 3(a). However, the VO made no specific allowance for the frontage because the nature of the building was reflected in the rent.

Finally, it was observed that no addition had been made for the existence of air-conditioning in the property, so a value was attributed to that using the statutory decapitalisation rate of 5.5% (see Appendix VI – Copy of Valuation, p27).

Through negotiation and research I was able to agree a reduction in RV from £19,500 to £12,250. I discussed this with my supervisor who was happy with my recommendation that the revised figure represented a fair assessment of the property.

Client reporting

I informed the client that I believed the original RV of £19,500 to be incorrect and over-assessed. Furthermore, following discussions with the VO I had agreed a revised and more accurate figure of £12,250 with effect from 1 April 2000. I felt that this was a more accurate and true reflection of the RV, and that I had therefore achieved the client's objectives.

I did not recommend that the appeal be taken to Valuation Tribunal (VT) because I felt that the revised RV was a fair assessment in light of the facts, and I did not believe that any new evidence would arise that would support a lower RV. It was my opinion that had the case been taken to VT, our chances of obtaining a favourable decision would have been slim. I advised that it would not be financially beneficial in relation to the level of risk and resulting rate savings (if any) that would be obtained.

I produced a report detailing the negotiations and my recommendations, and provided the client with a copy of our valuation. I also included a rate budget planner showing their revised liability and the amount we had saved them (see Appendix VII – not reproduced here). I advised the client that I thought the revised RV represented a fair assessment of the property and recommended their acceptance, which they provided in due course. Once this was obtained I was able to return the Agreement Forms to the VO, after checking all the details were correct.

Reflective analysis

Overall, I felt I had achieved my client's objective by obtaining them a fair assessment of the property, and minimising their business rates liability. By agreeing the revised RV, I saved the client £12,337 over the five-year period of the 2000 Rating List.

This case was a good learning process for me because there were many different areas that I had to examine in order to achieve the reduction. I needed to look at the specific issues that were unique to the property, such as the zoning approach adopted by the VO, and analysing the passing rent on the property. At the same time, I also needed to be able to look at the wider picture by examining what the comparable evidence was showing, and the state of the retail market generally. By doing both these things, I felt I was able to achieve the client's objective and obtain the correct RV.

There were elements of the project that were difficult, such as the fact that many of the appeals in the immediate vicinity had already been settled, thus making it hard to lower the tone to the same basis as the passing rent. This also meant that the VO had strong evidence to support their arguments. However, I was also aware how important the passing rent was in establishing the RV, and was able to emphasise this fact.

I also felt that the project showed me the importance of communicating effectively with colleagues and my supervisor. Due to my close involvement in the project, it would have been easy to concentrate on just one factor, such as the rent, however, by keeping my supervisor updated at all stages of the negotiations, I was able to ensure that all the main issues had been addressed.

I found the negotiations the most difficult part of the project, and it was a steep learning curve establishing what issues I should be emphasising in my discussions, and which issues I could be more accommodating on. Being involved in the negotiation process has also highlighted how important research is in establishing facts, but also that many aspects involve a degree of subjectivity and opinion. I appreciate that the more negotiations I undertake, the more my judgement and overall negotiation skills and confidence should improve.

Transferable skills

This project was useful to me in combining the statutory elements of rating with the practical realities faced in dealing with individual appeals. There were many different issues that arose which are common to a number of retail properties, and I consider that this case has provided me with a rounded view of rating, and how the different elements contribute to determining a property's RV. It has also provided me with experience in Landlord and Tenant and valuation issues, and illustrated how important a good overall knowledge of different areas is necessary – including and how they overlap in practice.

Appendices

The Appendices were: Appendix I – Location Plan & O.S. Extract, Appendix II – Description & Photographs, Appendix III – Lease Summary, Appendix IV – Rate Liability before Appeal, Appendix V – Schedule of Comparable Evidence, Appendix VI – Valuation, and Appendix VII – Summary of Rate Savings. Only Appendix VI is reproduced here.

Appendix VI – Valuation

Adopted main rate – £110

Floor/Description	Area (m²)	Factor	Rate (£/m²)	Value (£)
Ground – Zone A	62.74	A/1	£110.00	£6,901
Ground – Zone B	26.91	A/2	£55.00	£1,480
Ground – Office area (masked)	52.73	A/3[1]	£36.67	£1,934
Ground – Machine room	29.38	A/6[2]	£18.33	£539
Ground – Staff room	20.47	A/8	£13.75	£281
Ground – Store	27.28	A/10	£11.00	£300
Ground – Store (external)	4.18	A/20[3]	£5.50	£23
First – Store	45.21	A/16	£6.88	£311
Air-conditioning[4]	1			£547
Total				£12,316
			say	£12,250[5]

Note 1. Although this area was located in Zone B (and therefore should have been assessed at A/2) a relativity of A/3 was adopted to reflect that the area suffered from masking from the solid wall and ATM machine.

Note 2. This area had originally been assessed at a relativity of A/2 in the VO's original valuation. This was because it was felt that there should be an uplift for the office area. I was able to persuade the VO that this was not realistic given that there was no separate access to the area and therefore could not be separately let. The area was therefore treated as standard retail space with an allowance made for masking.

Note 3. This was assessed at A/20 to reflect that the stores were located externally and of much poorer quality.

Note 4 The air-conditioning was assessed by estimating the cost of installation, which in this case was estimated to be £9,940, and then it was devalued using the statutory decapitalisation rate of 5.5%.

Note 5. With effect from 1 April 2000, thereby maximising the rate savings for the client. (Also, rounding is to £250 in accordance with rating practice).

Illustration of issues/possible interview questions

Q Why was the 1995 appeal withdrawn?

Q When can a VON be served?

Q Does the fact the your firm was involved in the property's rent review constitute a conflict of interest?

Q What is the AVD?

Q Can you please clarify the rules on being able to backdate the savings?

Q As a general point, even if transition nullifies savings, there may still be a benefit in connection with the next revaluation in reducing the RV – why?

Q Can we examine the precise definition/assumptions of RV – what legislation covers this?

Q Are there any practical aspects which derive from this, and influence negotiations?

Q What is the position regarding banks' ability to recover VAT on the rent they may pay in practice (ie with leaseholds)?

Q Is there evidence of rentals for banks being less than A1 and other A2 uses?

Q Banking is an A2 use – can the VO consider A1 uses, or is there an issue regarding *rebus sic stantibus*?

Q What impact does a restrictive user clause have on the RV?

Q If looking more specifically at a property as a bank, why might zoning etc, and the usual characteristics sought by retailers not be as important?

Q Were there any implications regarding Part III of the Disability Discrimination Act 1995, effective from October 2004?

Q You referred to 'masking' – why exactly is an allowance warranted – do retailers/banks view this as a disadvantage?

Q What other disabilities may effect the rental value achievable for a retail unit?

Q Air conditioning was mentioned. Why is this separated out as an item, and not included in a rate per sq ft?

Q What differences were there in valuation assumptions, factors affecting rental value etc, between the rent review in 1999, and the basis of RV?

Q What is the IBRF?
Q What is a category 3(a)?
Q Why didn't you recommend taking the appeal to VT?
Q Why was the VO reluctant to accept the rent as the best evidence?

Chapter 4

Rent Review

This chapter examines a rent review undertaken on behalf of a landlord. This combines rental valuation with an understanding of lease terms. The report is contributed by Roland Shaw.

Introduction

This critical analysis examines my involvement in a rent review instruction in which my firm represented the landlord of an office property located within the town centre of Wilmslow, Cheshire. The instruction, completed in May 2002, was dealt with predominantly by myself, with supervision and guidance provided by my line manager at the time.

Landlord and Tenant is the field of commercial practice that I enjoy most, and in which I consider myself to be both enthusiastic and proficient. The management of this instruction demonstrated my valuation and negotiating skills, and required reasoning, detailed analysis and clear direction on my part.

The reflective analysis section discusses how my actions were justified in light of the successful completion of the rent review, and highlights the skills developed and experience gained.

The property

The first-floor premises at 76 Collin's Lane, Wilmslow, are occupied by Car-insure Ltd by virtue of a lease commencing 4 February 2000, and expiring 17 November 2006. The lease provides for the rent to be reviewed with effect from 19 November 2001, approximately five years prior to the expiration of the term. The passing rental of £64,200 pa, equating to £12.94 per sq ft, was established at the beginning of the lease. This includes the use of 16 dedicated on-site car parking spaces. My firm is also retained as managing agents for the property by the landlord, Lancaster Pension Fund, who acquired the freehold interest in 2000.

The premises benefit from B1 planning consent under the Town and Country Planning (Uses Classes) Order 1987, and are being used by the tenant solely for the purposes of office accommodation.

Initial issues

No conflicts of interest were registered when referencing those parties involved on the my firm's client management system.

The Terms of Engagement made clear our understanding of the client's instructions and expectations in relation to this review.

The client was asked to provide a copy of the occupational lease and associated documentation, which was assembled and sectioned within an ISO 9001 approved filing system.

Location

The subject property is prominently situated on Collin's Lane, in close proximity to the commercial and retail core of Wilmslow town centre.

Wilmslow is located approximately 11 miles south of Manchester city centre, and seven miles south west of Stockport. Wilmslow's proximity to the conurbation has resulted in its long-term development as a dormitory town for Manchester business people. The local population of circa 30,000 benefits from nearby high quality retail and leisure amenities, easy access to junctions 5 and 6 of the M56 motorway, and close proximity to Manchester International Airport (5.6 miles to the north west).

The commercial centre of the town is occupied by a wide variety of local, national and international operators. A location plan and a site plan are attached as Appendices 1 and 2 respectively (not reproduced here).

Accommodation and description

Dating from the early 1980s, 76 Collin's Lane is a three-storey reinforced concrete-framed building with red brick elevations, covered by a flat asphalt/slate roof. The property provides approximately 1,347 m^2 (14,500 sq ft) of lettable office accommodation. The interior of each demise features suspended acoustic-tiled ceilings with inset Category II fluorescent lighting, plastered and painted walls, carpet-tiled raised floors, gas-fired central heating, and single-glazed steel-framed windows.

The accommodation is open-plan in design, but also includes a separate kitchen area and a number of cellular offices and meeting rooms to the perimeter of the floor plate. With the exception of the core area, which includes the stairwell and a single eight-person passenger lift, the subject demise comprises the whole of the building's first floor.

The subject premises were measured in accordance with the RICS Code of Measuring Practice (5th ed), on a net internal basis: first floor premises, 453.10 m^2 (4,877 sq ft). However, it was noted that a floor plan and dimensions included within the lease documentation advised the floor area to be 4,962 sq ft. The floor to ceiling height was measured to be 2.75 m.

Under the terms of the lease, the tenant is entitled to the sole use of 16 on-site car parking spaces, reflecting a car-parking ratio of approximately 1:300 sq ft.

On the day of inspection, the premises were found to be in good and tenantable repair and condition throughout. Photographs of the subject property's external/internal design and condition are included within Appendix 3 (not reproduced here).

Client's objectives

Further to the initial telephone call, the receipt of a formal letter of instruction made clear the client's primary objectives. These can be summarised as follows:

- To provide a comprehensive rental valuation report, satisfying the criteria specified in the letter of instruction. In doing so, we were required to describe and address factors affecting the open market rental value of the subject demise. These were to include, for example, onerous lease terms, building condition and state of repair, and details of market transactions (open market lettings, rent reviews and lease renewals).
- To provide our opinion of rental value of the subject demise.
- To provide the best strategy to be employed in order to achieve the best rent.
- To create good rental evidence, which would impact at subsequent reviews within the same building.
- Ultimately, to secure a desirable investment profile and capital value for the property as a whole.

Key issues

These can be summarised as follows:

- To ascertain the flexibility of the lease's rent review clause and, in particular, to establish if time was of the essence in relation to the serving of notice.
- To ascertain the effect on rental value of other lease clauses – onerous, restrictive or otherwise.
- To establish as to whether the occupational lease benefited from the protection of the Landlord and Tenant Act 1954.
- To determine if rent review negotiations should be initiated with immediate effect, or tactically delayed until a more appropriate time.
- To ascertain if the tenant would be representing their own interests at review, or if professional representation would be sought.
- To determine as to whether and when a 'Calderbank' letter might be issued to the tenant.
- To determine as to whether it was in the landlord's interests to apply to the RICS for a third party referral if negotiations were later to prove unproductive.

Lease analysis

Having made a thorough examination of the lease and compiled a summary of the document (see Appendix 4, p37), it became clear that the majority of the covenants/clauses featured standard content, and were drawn on flexible terms. For example, none of the alienation, permitted user, repair, alteration or rent review clauses were considered restrictive or onerous in any way. It was noted that the lease had been drawn on effective full repairing and insuring terms.

However, what was regarded to be of significance was the fact that the lease had been contracted out of the Landlord and Tenant Act 1954. Given this information, it was clear that the tenant had no statutory right to renew the lease upon the expiration of the initial term in 2006. It was decided that this factor, allied to the short duration of the unexpired term, had the potential to impact on the rental value of the subject premises.

Market commentary and rental evidence

The Wilmslow office property market forms part of the broader south Manchester conurbation. The town is a well-established commercial centre and has in recent times re-emerged as an office location, together with the area immediately surrounding Manchester Airport. Modern road links and its proximity to the airport have contributed to the town's perceived desirability. The buoyancy of the town centre office market is also reflected by the continued proliferation of new office buildings to the fringes of Wilmslow, where developers have been eager to capitalise on a local market that achieves rents in excess of £17 per sq ft. However, it should be noted that these new-build office premises feature more flexible open-plan interiors and, typically benefit from superior specifications, improved accessibility and generous provision of on-site car parking.

Due to the depth and range of rental evidence available at Wilmslow, I was able to select and scrutinise only those examples that were truly comparable in terms of the size, specification and age of property, and also the occupational lease effective in each instance. Analysis of the evidence collated provided a range of rental values between circa £11 per sq ft and circa £15 per sq ft. Appendix 5 provides full details (see p41).

The two most recent transactions at 76 Collin's Lane were those involving the first-floor subject demise and the second-floor demise. The ground floor demise is occupied under a lease agreed in 1998. For details relating to the ground and second floor premises, refer to Appendix 5.

The letting of the second-floor accommodation to Sports Management Ltd, 13 months prior to the effective date of the subject's rent review, was deemed the closest comparable upon which to base our rental valuation. For the open market lettings of both the subject demise and the second floor demise, the respective leases had been contracted out of the Landlord and Tenant Act 1954. The need to reflect the impact of exclusion of security, coupled with a lack of representation by other market evidence with respect to this particular issue, meant that to do otherwise would have been misguided.

Rental valuation

Taking the comparable evidence into consideration, it was my opinion that the subject demise was under-rented, and that a significant uplift at this review was justified.

Adopting the rent achieved by the open market letting of the subject's second floor premises (ie £14 per sq ft), the rental value for the subject demise was calculated as follows:

4,962 sq ft @ £14 per sq ft = £69,468, say £69,500

Review strategy and serving notice

After reporting our findings to the client and, subsequently receiving instructions to proceed, a Notice of Rent Review (see Appendix 6 – not reproduced here) was served on the tenant via recorded delivery. So as to be admissible to third party at a later date if required, the notice was prepared as an open letter. The notice clearly identified the subject premises, the date of the lease and the date of the review being triggered. Also identified were the original parties of the lease as well as the current landlord and tenant.

As is also common practice, the notice stipulated the new rent proposed by the landlord. This, however, was not an actual requirement of the lease. Finally, the document is signed on behalf of the landlord by my firm.

It was decided to trigger the review with immediate effect by proposing a new rent higher than the market rental value calculated above. We believed this would allow scope for negotiation with the tenant's representative, as we expected to make downward adjustment of our initial demands in order to reach a settlement. Accordingly, the new rent proposed by the notice was based on a net internal floor area of 4,962 sq ft and £15 per sq ft – ie £74,500 pa.

The negotiation

During subsequent negotiations, which included telephone discussions, written correspondence and personal meetings with the tenant's representative, the following issues were discussed and challenged.

Floor area

We were able to argue successfully that our floor area was supported by the lease and the heads of terms relating to the original letting of the premises in February 2000. The tenant's representative agreed that we observe the floor area specified by the lease.

Rental evidence

The tenant's agent, while being fully aware of the recent letting of the second-floor premises at the same property, remained determined to cite examples of other lettings in the locality that demonstrated lower rental values in the region of £12 per sq ft. The agent was at first insistent that the review would have to be settled at a nil increase, or referred to third party for determination.

Upon closer examination of the agent's comparable evidence, it became clear that the lower rents corresponded to inferior office premises, or accommodation situated in less prestigious areas of the town. For example, the majority of the evidence related to office space located above retail units along the high street, and not purpose-built offices of a standard comparable to the subject property. Also, those purpose-built premises did not feature internal specifications of a comparable standard – such as lifts and raised floors were not included.

Exclusion of security

The tenant's agent was initially keen to highlight the implications of the subject's occupational lease being contracted out of the Landlord and Tenant Act 1954.

However, because no recent precedent existed other than the lease taken by Sports Management Ltd on the second floor of the same property, the tenant's agent decided to withdraw the issue from further negotiations.

Calderbank letter

As negotiations had become protracted and frustration was being felt by the client due to the tactics used by the tenant's representative, the option of serving an open offer letter on the basis of 'without prejudice save as to costs' on the tenant became increasingly favourable. However, we were able to prompt the tenant's representative by advising that such a letter would be served imminently, and that an application would be made to the RICS for the appointment of an arbitrator. Consequently, and due to the strength of the landlord's position, the tenant relented and intimated that the review could be concluded if the landlord was prepared to accept a new rent based on a compromise of £14 per sq ft.

Resolution

Accordingly, the parties verbally agreed a new rent of £69,500 pa. Upon receiving our client's instructions, the relevant Memorandum of Rent Review (see Appendix 7 – not reproduced here) was prepared in duplicate, and despatched to the tenant's representative for signature. The copies of the Memorandum were completed in due course. The fee invoice was raised and issued to the client in accordance with the original instruction.

Reflective analysis and conclusion

I consider that the landlord achieving a rental uplift of £5,300 at this review vindicated my actions. This new rent has secured a substantial increase in the total rent to be received from the five years remaining until lease expiry – a sum of £26,500. This review has also provided improved evidence for the future rent review of the ground-floor premises within the same building.

The instruction was a success in terms of meeting all the client's primary objectives. Principally, the review has resulted in elevating the investment profile of the property as a whole. However, it was recommended that the investment value would be further enhanced by ensuring that future lettings within the property were made subject to leases protected by the Landlord and Tenant Act 1954.

Throughout my involvement in this rent review, I have been able to develop both my technical knowledge and practical expertise. Accordingly, I now fully appreciate the importance of the following:

- Using effective negotiation skills.
- Drawing on the experience and advice of colleagues.
- Referring to industry literature and additional sources of market information.
- Having a clear understanding of the client's requirements and objectives from the outset.
- Acting in a professional manner at all times.

As a result of my involvement in this case, I now have a better understanding of the long-term and further-reaching benefits that can emanate from properly implemented estate management.

On reflection, I feel certain that the experience and confidence acquired during the management of this instruction will prove invaluable in future work undertaken within commercial real estate practice.

Appendices

The Appendices were: Appendix 1 – Location Plan (1:100,000), Appendix 2 – Site Plan (1:1,250), Appendix 3 – Photographs of the Subject Property, Appendix 4 – Lease Précis, Appendix 5 – Comparable Evidence, Appendix 6 – Copy of Notice of Rent Review, and Appendix 7 – Copy of Memorandum of Rent Review. Appendices 4 and 5 are reproduced below.

Appendix 4 – Lease précis

Landlord

Lancaster Pension Fund.

Tenant

Car-Insure Ltd.

Demise

First floor, 76 Collin's Lane, Wilmslow, Cheshire.

Dated

4 February 2000.

Contractual term

For a term commencing on 4 February 2000 and expiring on 17 November 2006.

Exclusion of security

An order was issued by Manchester County Court on 1 February 2000 to exclude sections 24–28 of the Landlord and Tenant Act 1954 (Part II) – ie this lease is outside the Act, and the tenant has no statutory right to renew at the termination of the lease.

Rent

£64,200.

Interest

4% above the base rate in force at The Royal Bank of Scotland plc.

Permitted user

The premises may be used as offices within Class B1 of the Town and Country Planning (Use Classes) Order 1987. The tenant shall not use the premises for any purpose except the permitted use.

Alienation

The tenant shall not assign, underlet, charge, part with or share possession or occupation of all or any part of the premises, nor hold the premises on trust unless expressly permitted to do so by the landlord.

The landlord shall not unreasonably withhold consent to a legal charge of the whole of the premises.

The landlord shall not unreasonably withhold consent to an assignment of the whole of the premises, although the following conditions need to be satisfied:

- The assignee is not a 'group company' or a 'connected person'.
- The assignee is of sufficient financial standing.
- The assignor enters into an Authorised Guarantee Agreement with the landlord, in such terms as the landlord may reasonably require.
- The assignee must provide an acceptable guarantor if one is required by the landlord.

The landlord shall not unreasonably withhold consent to an underletting of the premises as a whole where the following conditions are satisfied:

- The undertenant observes and performs all tenant's covenants in the lease.
- The undertenant must provide an acceptable guarantor if one is required by the landlord.
- No fine or premium is taken for the grant of the underlease.
- The rent payable under the underlease remains unchanged from that due under the head lease.
- The form of the underlease has been approved in writing by the landlord (approval not to be unreasonably withheld).
- The tenant shall enforce the terms, conditions and provisions of any underlease, and shall not accept a surrender of part only of the underlet premises.
- The tenant shall not, without consent, vary the terms of the underlease, accept a surrender of the underlet premises or agree any review of the rent under any underlease.

Rent review

The basic rent is due for review on 19 November 2001, five years prior to the expiration of the lease. Rent reviews are upwards only and time is not of the essence.

Assumptions are as follows:

- Willing landlord and willing tenant.
- A letting of the whole of the premises in the open market.
- Vacant possession.
- As per the terms of this lease, excepting stipulations in the lease as to the amount of rent.
- Without any fine or premium being taken.
- To be valued as for a new lease equal in duration to the term unexpired or for a term of seven years (whichever is longer).
- All the covenants in the lease have been complied with.
- If the premises or the building, or any part of either, have been destroyed or otherwise damaged, then they have been fully restored.
- The premises comply with all legal obligations and may be used lawfully for the purpose permitted by the lease.
- The premises are fit and ready for immediate occupation and use.

Disregards are as follows:

- Occupation of the tenant.
- Goodwill.
- Any effect of the tenant's fixtures or fittings.
- Any improvements to the premises carried out by and at the expense of the tenant.
- Any actual or potential obligation on the tenant or any undertenant to reinstate alterations or additions to the premises.
- Any temporary works of construction, demolition, alteration or repair being carried out or near the building.
- Any work carried out to the premises which diminishes their rental value at the relevant review date.

In respect of third party determination:

- If the landlord and tenant are unable to reach agreement as to the level of the revised rent three months before the relevant review date, then either party may require the matter to be determined by an impartial rent review surveyor. The rent review surveyor can be appointed between the two parties or by referral to the President of the RICS.
- The rent review surveyor shall act as an arbitrator in accordance with the Arbitration Act 1996, and his award shall be binding on the parties.
- The rent review surveyor has the power to make a provisional award before the final award is determined.
- If the Open Market Rent is not agreed or determined until after the relevant review date, the tenant shall continue to pay the basic rent until the day on which the revised rent is agreed. The tenant will then become liable to pay the amount of any increase for the interim, together with interest calculated at 2% below the interest rate for the period.

Repairs

The tenant shall keep the premises in good and substantial repair and condition throughout the term.

Decorating

The tenant shall, three months preceding the expiration of the term, redecorate the premises to a high standard, observing a colour scheme that has first been awarded the landlord's consent (such consent shall not be unreasonably withheld).

Alterations

The tenant may carry out alterations, additions or improvements to the premises which are wholly within the premises and do not affect any part of the exterior or structure of the building where:

- The tenant has submitted detailed plans and specifications (in duplicate) showing the works proposed.
- The tenant has given to the landlord such covenants relating to the carrying out of the works as the landlord may reasonably require.
- The tenant has received consent to the works (such consent not to be unreasonably withheld).

Landlord's costs

The tenant shall pay to the landlord, within 10 business days of a written demand, all reasonable and proper costs, expenses, losses and liabilities incurred by the landlord in connection with:

- Any breach by the tenant of its covenants or obligations and/or any enforcement of those covenants and obligations by the landlord.
- The preparation and service of any notice under section 146 or 147 of the Law of Property Act 1925.
- The preparation and service of any notice concerning a tenant's breach of the repair and decoration covenant, or any schedule of dilapidations served during the term or within three months after determination.

Yield up

At the expiration or sooner determination of the term, the tenant shall yield up the premises to the landlord with vacant possession and in a state of repair, condition and decoration consistent with the proper performance of the tenant's covenants under the lease.

Service charge

The service costs are all those proper and reasonable costs and expenses incurred by the landlord in the operation and management of the building and the provision of services to the building.

The landlord will notify the tenant before or during any service charge period of the estimated service charge payable by the tenant.

The landlord may, acting reasonably, revise the estimated service charge, and the tenant shall pay any additional demand within 10 business days of receipt.

The proportion of service charge is not a fixed percentage of the annual cost and is liable to vary from one service charge period to the next.

Car parking

The tenant is entitled to the exclusive use of 16 car parking spaces.

Appendix 5 – Comparable evidence

Ground floor

Landers hold a lease of seven years, 11 months and 22 days duration, with effect from February 1998. The premises comprise the whole of the ground floor and extend to 4,638 sq ft. The passing rent of £41,640 pa, equating to £8.98 per sq ft, includes the use of 16 on-site car parking spaces. The lease, like that relating to the subject demise, is excluded from sections 24–28 of the Landlord and Tenant Act 1954.

Second floor

Sports Management Ltd took a five year lease with effect from 29 September 2000, on the whole of the second floor, comprising 4,750 sq ft. The passing rent of £66,500 pa, analysing at £14 per sq ft, includes 16 on-site car parking spaces. As with the above, the lease has been contracted out of the Landlord and Tenant Act 1954, and the tenant will have no automatic right to renewal upon the expiration of the term in September 2005.

Other comparable properties in Wilmslow

Among examples of new-build satellite premises, The Worcester Group had recently taken occupation of 12,345 sq ft at Medway Court, Wilmslow, by virtue of a 15-year lease and a rent equating to £17 per sq ft. Also, Grinsteads had taken 3,204 sq ft of office space at Malverley House, Nestle Park, Wilmslow, on a 20-year lease with an option to break at year 10, at a rent analysing as £16.95 per sq ft.

Elsewhere, examples of lesser rents for second-hand office premises included Nottingham House, Plough Street, Wilmslow, dating from the early 1980s. Here, Garden's Finance had recently occupied the whole of the four-storey property, comprising 23,360 sq ft, on a 10-year lease, at a rent of £15 per sq ft. Dating from the same period, Malone Court at Dove Street is a four-storey development, comprising 23,000 sq ft, which generally achieves rents of £14 per sq ft on 10-year leases.

Closer to the town centre, passing rents typically become lower. For example, Antex Ltd had taken a 788 sq ft office suite at 19–21 Collin's Lane on a three-year lease, at a rent analysing as £13 per sq ft. Suite 11, Blumenau House, 41 Collin's Lane was let in December 2000 to M C Jones for a five-year term. The rent equates to £13 per sq ft, based on a floor area of 7,696 sq ft. Also at the same address, Suite 10 was let to Creative Designs Ltd in January 2001 for a five-year term, at a rent equating to £11.90 per sq ft on a floor area of 7,976 sq ft. Most recently, Virana Ltd had taken a five-year lease on Suite 8, at a rent breaking back to £13.50 per sq ft, based on 7,976 sq ft. The accommodation provided at these properties is of lower grade specification.

It should be noted that the majority of the rents highlighted above include on-site car parking at a ratio comparable to that observed with the subject demise – i.e. approximately one space per 300 sq ft.

Illustration of issues/possible interview questions

Q Can you remind us of the construction of the property?

Q Were there any particular maintenance defects?

Q What would you say were the attractions of the property to prospective tenants regarding specification?

Q What are service charges per sq ft for the subject property?

Q How did this compare to the comparable properties, and in your wider experience, does a service charge differential become a particular negotiating point at rent review?

Q Regarding the comparable evidence, you indicated that Virana Ltd had taken a five-year lease on Suite 8, at a rent breaking back to £13.50 per sq ft. Presumably this was broken back because of a rent-free period – how did you undertake the calculation?

Q The area for that letting was approximately 8,000 sq ft, whereas your property was approximately 5,000 sq ft. Is there a valuation issue in connection with different sizes?

Q What is meant by the 'heirarchy of evidence'?

Q You referred to contracted out tenancies having an effect on rental value – what factors influence the extent of any adjustment which might be made?

Q In the appendix, you stated the disregards of the tenant's occupation and any goodwill it has created. Why is this necessary?

Q Why is it necessary to have a disregard of goodwill, as well as tenant's occupation?

Q Can you explain in more detail the role of the arbitrator?

Q How does this differ from that of an independent expert?

Q What in your experience is preferred by landlords and tenants out of arbitrators and experts, and are there any particular factors this depends on?

Q Can you explain in more detail how the rent review enhanced the capital value of the property?

Chapter 5

Rent Review – Further Illustrations

This chapter includes extracts from *Estates Gazette*'s 'University Links' feature which reported on workshop training events run at Heriot-Watt University in Edinburgh, and supplementary information relating to the event. This was a joint initiative between RICS, GVA Grimley, and Heriot-Watt University, with investors Haslemere and New Edinburgh also being involved. Additionally, rent review case law illustrations are provided.

Rent review involves many tactical, as well as technical, considerations. Surveyors' work in each case needs to reflect the varying circumstances of landlords, and how this affects the strategies deployed in rent review and other work.

Effect of rent review on capital value

The amount of rent agreed at rent review will influence the capital value of the property, and the review therefore provides an opportunity for a landlord/investor to increase capital value.

Property investors especially will be conscious of the capital value of their property interests. This is, of course, important if they wish to sell the property, as they will realise more profit.

Increases in capital value will also facilitate refinancing — many investors rely on loans from banks to finance their property acquisitions/investments. This will provide funds for the acquisition of new properties or to generally assist cash flow. Some investors, despite not looking to sell or needing finance, will still be conscious of how capital value affects portfolio/fund performance figures.

Landlords such as local authorities and the corporate sector will often be less conscious of capital value, and rather, instead, focus on the level of rent/revenue that can support the operational side of their business. This will have a particular influence on letting strategy, with flexible terms helping to attract tenants and secure the best rents. In contrast, private investors will be particularly concerned about income security, achieved through longer leases, and tenants of good financial standing (known as covenant strength).

Effect of the review on other rent reviews

Where a landlord has other tenants in the same building or in nearby buildings, the amount of rent agreed on the subject rent review will influence the rent achievable on other rent reviews (and also statutory lease renewals in England), as the subject rent review settlement will provide comparable rental evidence. The surveyor therefore needs to be aware of the extent of the landlord's interests in the area, and details of tenants, rents, impending rent reviews and so on.

Influence of nearby transactions

There may be lettings, rent reviews and lease renewals taking place within the building or nearby (including in other landlords' ownership) which could provide beneficial comparable rental evidence for the subject rent review. It may be helpful to delay the subject rent review (if possible) in order that other lettings and rent reviews can be concluded, and then presented as comparable rental evidence to the tenant. Another tactic could be to commence a rent review against a particular tenant who is likely to offer less resistance than others, with the evidence being drawn on for reviews implemented subsequently with the tenants likely to negotiate more strongly.

Scope for higher-value uses

When instructed in respect of a rent review, the surveyor may identify scope for a higher-value alternative use than the existing use to which the property is being put. This could include a change of use from general retail (A1 England, class 1 Scotland), to café/restaurant/takeaway use (A3 England, class 3 Scotland) or scope to refurbish/redevelop the building, such as converting it from secondary offices to residential apartments.

Tenant's compliance with lease terms

When inspecting the property at rent review, consideration also has to be given to the tenant's compliance with lease terms. Although the landlord may be prepared to turn a blind eye to some indiscretions by the tenant, discovery of non-compliance may still help to conclude the overall negotiations in the landlord's favour.

All-round surveying skills

Although a rent review is part of landlord and tenant work, the above points show that the surveyor's role incorporates rental valuation; capital valuation and investment considerations (and possibly finance issues); estate management regarding the compliance by the tenant with the terms of the lease; and planning and development regarding alternative uses.

Indeed, it is crucial for surveyors to be capable of lateral thinking and entrepreneurial thought processes in line with their clients' specific business interests. Surveyors' performance in rent reviews, lease renewals, agency and management work is a key, and sometimes underestimated, contributor to investment returns.

Progressing the review

The basic processes involved in rent review are listed below.

Taking instructions

The surveyor must be able to act for the landlord without having a 'conflict of interest'. If, for example, the surveyor had already been appointed by the tenant in respect of the rent review, he could not also act for the landlord. Terms of business (the contract between the surveyor and the client landlord) must be agreed in writing before the surveyor can proceed to act for the landlord. The surveyor may have had to initially pitch for business, and ensure that fees were competitive, in addition to expressing their expertise and track record.

Assessing the client's objectives

An initial meeting with the client landlord will establish his or her overall objectives. As well as the primary objective of securing the highest possible rent at rent review, as mentioned above, the landlord may have plans to sell the property, or may wish to consider refinancing.

Property inspection

In addition to recording factual details about the property and undertaking measurements, the surveyor is looking for factors affecting value which can help justify the new rent proposed to the tenant, and which will be raised in negotiations with the tenant's surveyor. On site, details will be recorded of other properties to let, new developments in the area, and so on.

Lease analysis

This will be important because the lease stipulates the basis of valuation to be adopted in the review, ie a market rent, and landlord's and tenant's obligations, some of which may have a significant effect on the rental valuation and rent review settlement achievable (such as restrictive user and alienation provisions, or onerous repair/service charge provisions). The lease will also contain assumptions and disregards; examples of standard provisions include the property being assumed to be vacant and to let, the occupation of the actual tenant being disregarded, and the value of any improvements undertaken by the tenant being disregarded.

Valuation

The rental valuation undertaken by the landlord's surveyor will be based on comparable rental transactions, such as lettings, rent reviews and lease renewals. These will sometimes be in the same building or nearby buildings, and possibly on similar lease terms, therefore making the surveyor's task relatively easy. Where there is no immediate evidence, surveyors will have to make adjustments to the valuation for factors such as location, prominence, specification, age, size, use, etc. Some transactions will not be representative of market value; examples include surrenders and renewals (which can reflect particular requirements of one or both of the parties), or assignments where a tenant is desperate to relinquish lease liability and property overheads.

Instigating the rent review, and negotiations

The landlord's surveyor will serve notice on the tenant to instigate the rent review in accordance with the provisions set out in the lease. The landlord's surveyor and the tenant's surveyor will then negotiate the rent. Early tactics could involve stating a relatively high rent in order to provide more room for negotiating, and perhaps a better position if the rent was determined by a third party (which will be an arbitrator or independent expert depending on the terms of the lease). The parties are also likely to be selective about the comparable evidence submitted to the other side, with the landlord supplying only the best rents, for example. Occasionally, the availability of a break clause could see a tenant threaten to leave if the new rent was not agreed at a certain figure.

Delay

A tenant may wish to delay settlement in order to put off paying the increased rent due once the revised rent is eventually agreed. However, as well as the possibility of the settlement rent being higher because better comparable evidence emerges for the landlord (even though the valuation date does not change), there may be a provision in the lease regarding interest on the late determination of the rent review, such as 4% above base rate, backdated to the date of the review.

The scope to delay instigating the rent review is due to 'time not being of the essence' in most leases. Where a landlord is delaying, and has not served a rent notice, despite the effective day of the rent review having passed, the tenant may be able to serve a notice on the landlord requiring the review to be instigated (unless the tenant can instigate the rent review, but this is unusual).

Landlords and tenants may, of course, wish to expedite a settlement, such as when the tenant requires certainty as to the rental/property overheads, or the landlord wishes to sell the property. Occasionally, the parties may agree to revise other lease terms, or even agree a new lease entirely.

Third-party determination

Tactics will also be available in connection with the possibility of third-party determination, and the service of Calderbank letters. The 'Mainly for Students' articles of 20 April and 18 March 2002 provide more detail on this, and on the roles of arbitrators and independent experts.

Case law illustration

As an illustration of the need to keep up to date with case law, an extract is set out below of two cases covered by *Estates Gazette*'s 'Mainly for Students' series, based on a landlord and tenant update hosted by RICS West Midlands CPD Foundation and Advantage West Midlands in January 2004.

As surveyors become involved in more detailed landlord and tenant work for higher-value property interests, it is increasingly important to be up to date with case law. As well as helping enhance credibility and the range of arguments put forward within negotiations, the ability to draw on points established in previous cases can help successful settlements to be reached, and best protect a client's interests.

Rent review treatment of premiums

Christopher Hancock of Wragge & Co LLP began by examining whether a premium for the grant of a lease is part of the rent, or is key money (a payment simply to secure occupation. This is a current issue in the retail market, particularly between tenant supermarket operators and their investor landlords, with substantial premiums often reflecting increasingly restrictive planning policies, and the reduced number of new sites.

In *BLCT (13096) Ltd* v *J Sainsbury plc*, reported in the Court of Appeal: [2003] All ER(D)03 (July), the relevant part of the rent review clause provided that 'market rent' meant 'the higher of the yearly rental aggregate of the yearly rents at which the demised premises might reasonably be expected to be let as a whole...by a willing lessor to a willing lessee with vacant possession without any premium in the open market at the relevant review date...'.

Both parties agreed that an Asda store was the nearest comparable, but could not agree how to analyse offers made in July 1999 from Safeway at £20.48 per sq ft and Asda at £18 per sq ft plus a premium of £3m. This resulted in a difference of £5 per sq ft and would have given a revised figure of £23 per sq ft if rentalised. The premium was not expressed as being for any particular purpose.

The arbitrator, Graham Chase, treated the transaction as if no premium had been paid, finding that it was 'key money', with no evidence that it equated to a rental payment. He took the view that there was no single approach to the treatment of premiums and also that they might be amortised at different rates.

Details of another case, *Safeway and Legal & General*, are not presently in the public domain, but Legal & General succeeded in persuading the arbitrator to take into account a £3m premium paid by another supermarket group at a different location.

These are essentially valuation, not legal, points, with the Sainsbury's Court of Appeal ruling being a refusal to grant leave to appeal. More precise consideration may still, however, be necessary in lease drafting.

If, as a simple example, a key comparable for a rent review of a shop is a five-year lease at £200,000 pa (say £200 per sq ft zone A) and a £250,000 premium is payable, the landlord's favoured analysis could be a market rent of £257,700/£257 per sq ft (£200,000 rent plus £57,700 annual equivalent of premium, calculated as £250,000, YP five years at 5% = 4.33).

The tenant's favoured analysis is £200,000, treating the premium as key money. If other comparables generally show the market level to be £200 per sq ft zone A, the premium is likely to be regarded as key money. If other comparables indicate £257 per sq ft, the premium is more likely to be regarded as part of a tenant's rental bid. The valuation difficulties arise when comparable evidence is limited, as in the Sainsbury's case, and neither the market rent for the premises, nor the general tone of values can be established with accuracy.

Rent review length of term

Another recent issue is whether, at review, the hypothetical term to be valued equates to the remainder of the existing term, or is assumed to be the full term of the original lease. Neuberger J considered this in *Canary Wharf Investments (Three) v Telegraph Group Ltd* [2003] EWHC1575 (Chancery).

The length of the term affects rental value. As an example, for large office premises, a 15-year term may command the highest rent, but a 25-year term would be considered an onerous commitment, and a five-year term would be considered to offer insufficient security. In both the latter cases, rental value would be depressed. Sometimes, however, shorter leases will command higher rents owing to the greater flexibility afforded to the tenant.

In the *Telegraph* case, the landlord argued that the hypothetical term of 25 years ran from the commencement of the lease in 1992, rather than the date of the review in 2002, and therefore provided a 15-year term upon which rental value should be assessed. The tenant's preference was naturally for the 25-year assumption.

The decision in the case turned upon the interpretation of the lease provisions, and is the first known reported decision that goes against the often-perceived norm of applying the presumption of reality: valuing the residue of the original term at review.

Neuberger J found that the presumption of reality, always a good starting point in the interpretation of rent review clauses in the absence of clear wording requiring the parties to depart from reality, is something that should not be mechanistically applied. Neuberger J noted that in the most important cases where the presumption of reality had been relied upon, this was generally because there was insufficient provision in the lease for the court to be able to do otherwise. Taking account of the overall construction of this particular lease, he did not find that to be the position. Neither did he find it necessarily unfair that the term date should commence, for a term of 25 years, on the review date for the purpose of review. He took the view that, if this was the most natural meaning of the wording, it could not be said to have disadvantaged the landlord or the tenant as such, since a 25-year term presumably suited both of them at grant, and neither could foresee how values might be affected over the term as a whole.

The case might be a one-off, turning on its own special facts. But, whether or not the case indicates some relaxation in the circumstances in which the courts might seek to apply the presumption of reality, it does at least serve as a reminder that one should not readily presume that, at review, the term to be valued is either the remainder of the initial term, or a term equivalent to the original term, commencing on the rent review date.

Chapter 6

Lease Renewal

This chapter examines a lease renewal of retail premises undertaken on behalf of a landlord. The report is is contributed by Shrav Khera.

Introduction

My organisation's primary purpose is to run an efficient operational business, and a secondary purpose is to optimise returns from the commercial estate. I work as part of a small team of surveyors responsible for approximately 1,000 properties. My portfolio comprises a diverse mix of commercial, industrial and residential premises in North-West London. My primary duties encompass general estate management, rent reviews and lease renewals.

In compliance with RICS requirements I have obtained the organisation's/ client's permission to disclose the following information in this report.

This critical analysis examines the renewal of a lease for a retail premises, located within a retail parade in London, where I acted for my landlord organisation.

I have chosen this project because it has demonstrated my skills in all but one of my chosen competencies, namely Valuation, Property Inspection, Drawings, Landlord and Tenant and Estate Management.

Aims

The aims from my client's point of view are:

- Maximise the rental income generated from the investment.
- Ensure that the property is maintained in a good state of repair.
- Sustain a good landlord and tenant relationship.
- Ensure compliance with health and safety standards.

The report addresses the following issues:

- Reading of the original lease, and reviewing the property file to identify the current lease terms and other issues.
- Detailed inspection of the property and location.
- Analysis of comparable evidence.
- Rental valuation of the subject property, and preparation of the Heads of Terms.
- Negotiation strategy, and negotiation with the tenant's qualified surveyor.
- Agreement and the instruction of solicitors.

Location and description

The property is located within the retail parade. This is situated mainly within a residential area, and there are two adjacent retail parades owned by a private landlord, as well as a mix of commercial uses in the wider area. The development of a supermarket in close proximity has resulted in voids arising within the other two parades) although my client's parade is currently fully let (see Appendices II and III – not reproduced here).

Transport links are good with the nearby mainline station and underground station, and numerous bus routes also service the local area.

The subject property comprises a single storey mid-terrace ground floor lock-up shop. The property is of solid masonry construction with a glass frontage, suspended ceilings and solid floors. Deliveries can be made to the rear by means of a right of way over the landlord's property.

Internally, the premises comprise a sales area with small ancillary storage which incorporates a kitchenette and wc at the rear.

Lease terms

In order to prepare my valuation, I first reviewed the lease. (See existing lease, Appendix VII – not reproduced here.) A summary of the principal terms is set out below:

Term:	Three years from 15 November 2000.
Rent:	£8,000 pa exclusive from 15 November 2000.
User:	The use is restricted to a ladies' and gentlemen's hairdressers.
Insurance:	Tenant to insure.
Repairs:	Tenant has full repairing obligations.
Alienation:	Tenant is able to assign the whole of the tenancy with the landlord's consent, which is not to be unreasonably withheld. There is an absolute covenant against the assignment of part, and sub-letting of the whole or part of the premises.
Alterations:	The tenant is entitled to make non-structural alterations with the written consent of the landlord, not to be unreasonably withheld. There is a complete bar on structural alterations.
Break clauses:	The term is subject to early termination by the landlord at any time upon six months prior written notice if required for the landlord's own purposes or for redevelopment. In exceptional circumstances, the term may be determined on 28 days' prior written notice, but only in the case of the premises being urgently required for urgent works or repairs to the landlord's operational property and where a certificate is first obtained from the appropriate government Minister.
Security of Tenure:	The lease is fully protected by the Landlord and Tenant Act 1954, Part II.

Property inspection

As part of my inspection, I carried out a detailed measurement of the premises to determine the Net Internal Area in accordance with RICS Code of Measuring Practice, 5th ed.

Due to the depth of the premises, the zoning method was adopted. I zoned the property on the basis of 6.1m zones and calculated the floor area in terms of Zone A (ITZA) as 42.38 m² (456 sq ft).

During my detailed inspection, I also checked covenant compliance in order to ascertain if there were any issues which needed to be raised, in particular the repairing covenant and Health and Safety matters. In my opinion, the premises appeared to be in good condition and I decided that a schedule of dilapidations was not required.

I also checked whether the tenant had undertaken any works which may be classed as tenant's improvements under the Landlord and Tenant Act 1927, and for which consent had been granted. These would need to be disregarded on renewal, although from the inspection, I noted that there were no such improvements.

During the inspection, I also observed that there were two vacant units within the adjacent parades, details of which were noted for further investigation.

Comparable evidence

The most obvious comparables to use were the landlord's adjacent properties situated within the parade. These properties have been let on similar lease terms and, significantly, all include the landlord's option to break the lease, similar lease terms and restrictive user provisions (see Appendix V – not reproduced here).

Although not subjected to the same pedestrian flow, there was a possibility that the two adjacent parades could provide rental evidence to support an rental increase for the subject property. Upon further investigation, I ascertained that both were owned by a private property company. Their surveyor confirmed that since the opening of the supermarket, they had struggled to maintain full occupancy – in some cases letting to unattractive covenants.

After further investigations, the most recent evidence confirmed a rate of £177.97 per m² (£16.53 per sq ft). This convinced me that the best level of evidence would be derived from the landlord's own properties within the parade. This was because of the similar location and lease terms, and because the level of demand within the parade was similar to that of the subject property.

Comparable evidence and rental valuation

Shops 4 and 8 were recently agreed at rent review, however I considered shops 4 and 7 to be the best comparable evidence since they were open market lettings.

I used the evidence for shop 7, due to the transaction date being more recent. Before completing my valuation, there were factors that needed to be considered which required an adjustment to the base figure. These were:

- Alienation rights: there was a prohibition on any form of assignment or subletting, and an adjustment of 2.5% was appropriate.
- Security of Tenure: the lease was excluded from sections 24–28 of the Landlord and Tenant Act 1954 Part II, therefore the tenant has no security of tenure, and an adjustment of 2.5% was made.
- Rent Review pattern: the property had a rent review after the third year of the term, and due to this difference an adjustment of 2.5% was applied.
- Time: there was a time lag between the letting and the valuation date, therefore a further 2.5% adjustment was made.

After careful deliberation with my supervisor, we decided to increase the comparable figure by 10% to reflect these differences. My valuation was therefore:

42.38 m²   × £235.96 m²
(456 sq ft)   (£21.92 sq ft)
             £9,999
say   £10,000 pax

Options

At this stage there were a number of alternatives:

Option 1: Do nothing

If the market rent was found to be considerably lower than the passing rent, the realistic option would be to maintain the current rent. However, the current rent was below market rent and therefore rental negotiations should commence without further delay.
Not recommended.

Option 2: Informal negotiations

This would have been an attractive option if the market evidence showed a slight increase, and the landlord and tenant relationship was good. However, after reviewing the file, and consulting with my supervisor, I concluded that it would be difficult to achieve an uplift in the rental by only informal negotiations.
Not recommended.

Option 3: Formal procedure

I may have considered serving a section 25 notice which opposed the grant of a new tenancy, as outlined within section 30 of the Landlord and Tenant Act 1954. However compensation would have to be paid to the tenant for disturbance, and there was no reason to seek possession. The landlord intended to grant a new tenancy.
Not recommended.

Option 4: Formally terminate the lease

The service of a section 25 notice would, through court directions, set a rigid timetable for both parties to agree the terms of the new lease.
Recommended.

A valuation report was compiled and submitted to my supervisor for approval, based on the comparable evidence. I suggested that terms should be served at £11,500 pa exclusive, with immediate service of the section 25 notice.

Lease renewal negotiations

Following the approval of the valuation, I prepared and served a non-opposing section 25 notice, sent by Royal Mail Special Delivery on 8 April 2003, bringing the lease to an end at the contractual lease end date of 14 November 2003.

The notice and the covering letter were copied to the landlord's solicitor. The counter-notice was received on 7 May 2003 and the originating application to court on 16 July 2003.

I instructed the landlord's solicitor to file an answer to the court, proposing that the landlord would be willing to grant a new three-year tenancy at £11,500 pa exclusive. I also requested that an interim application be made at an unspecified level. All the other terms were to be as the existing document, subject to the modernisation of the lease terms to the landlord's current standard form.

At this point I forwarded the proposed Heads of Terms to the tenant, who appointed a chartered surveyor.

Subsequent to the tenant's solicitor's application, the court gave notice that a Directions Hearing was listed for the 12 September 2003. The landlord's solicitors requested my instructions. The options were:

- Request adjournment to allow a reasonable period for negotiations to commence.
- Agree directions to set down a formal timetable.
- As no response had been received to my proposed Heads of Terms, I instructed the landlord's solicitors to seek an initial adjournment of two months in order to allow negotiations to proceed. A revised date for the directions hearing of 14 November 2003 was received.

It was some four weeks until a reply was received from the tenant's appointed surveyor. As expected, the proposed rental was rejected, and counter-offer was made at £9,250 pa exclusive. All other proposed terms were however agreed.

At this stage I submitted my floor areas and comparable evidence to assist with negotiations. The floor areas were agreed.

I argued with the tenant's surveyor that the comparable evidence should be increased based on the following issues:

- No alienation rights.
- Exclusions from sections 24–28 of the Landlord and Tenant Act 1954.
- The differences in rent review patterns.
- Time elapsed from the letting to the valuation date.

The opposing surveyor disagreed with my quoted rent of £271.35/m² (£25.22/sq ft). He argued that no adjustments were applicable as it was the landlord's policy only to grant excluded leases, and prospective tenants would have offered the same rent if protected.

During negotiations new evidence occurred with the open market letting of shop 2 in the parade.

Tenant's agent:	Professionally represented.
Term:	Six years from 6 August 2003.
Review pattern:	Three-yearly upwards only.
Rent:	£10,000 pa exclusive.
User:	Restricted to the retail sale of digital printing stationary and computer equipment.
Alienation:	Assignment/underletting not permitted.
Re-entry:	Landlord's re-entry on six months notice at any time, or on 28 days' notice for urgent repairs to the subject premises.
Accommodation:	Ground floor retail unit with ITZA of 38.93 m² (419 sq ft) at £256.87 per m² (£23.87 per sq ft).
Security of tenure:	Excluded from the Landlord and Tenant Act 1954, Part II.

The above comparable was submitted due to the property being similar in terms of location and size. The tenant's surveyor argued that the property was not comparable due to the varied lease terms. I emphasised that since the property was within the landlord's portfolio, and also was an open market letting, this did make the property comparable and relevant to our negotiations.

It was agreed that interim rent was not an issue. Therefore the new lease was to commence from the expiry of the previous lease – 15 November 2003. Furthermore, the tenant would pay the balance of rent on completion. However, should negotiations have slowed down I may have considered the issue of interim rent.

Legalities

At the 11 December hearing, court directions were approved by the court. I instructed the landlord's solicitors to serve a draft lease upon the tenant's solicitors, not including the rent. The tenant's solicitors agreed to the terms of the proposed lease. Following this, I revised the current situation of which I was presented with two options:

- Request a stay in proceedings to allow for negotiations to continue.
- Allow the current proceedings to progress to court.

I decided to instruct the landlord's solicitors to apply for a further stay in proceedings to allow enough time for an agreement to be reached. I felt it in my client's best interest to reach agreement before the court hearing.

The tenant's surveyor suggested the following terms:

Term:	Six years from 15 November 2003, subject to a three-yearly upward or downward rent review pattern.
Rent:	£10,000 pa exclusive 15 November 2003.
User:	Restricted to an A1 use.

The tenant's surveyor agreed to the other terms of the lease. The options were to:

- Agree to the tenant's request and grant the lease as required.
- Apply to the court for directions.

I believe the tenant wanted to lengthen the term in order to improve the value of the lease, and provide greater security of tenure.

After further consultation with my supervisor, I decided that, due to the secondary location of the property, it would be more beneficial for the landlord to grant a longer term. Should the property have become vacant, it could result in no income being produced for a period of six to nine months due to its location.

The following terms were agreed:

Term:	Six years from 15 November 2003, subject to a three-yearly rent review pattern.
Rent:	£10,000 from 15 November 2003.
User:	Restricted to A1.

I considered that this agreement would continue the good landlord and tenant relationship with the tenant.

Since agreeing the terms of the lease, the engrossed lease has been submitted to the tenant's solicitors for execution. I have informed the landlord's litigation solicitors of the recent developments. Upon completion of the legal formalities, I will instruct the solicitors to discontinue court proceedings.

Critical appraisal

Overall, I feel that the project went well. However, I may have been inclined to adopt a different approach had progress been slower.

Instead of requesting a stay for negotiations to commence, I could have given instructions immediately to agree the revised directions. The implication of this would be to introduce a strict time scale which should be observed, and would speed up the negotiation process. The landlord would then have been able to apply for the Tenant's Application to be struck out if they failed to comply with the timetable.

The process of negotiations may be quickened by the service of a Part 36 Offer under the Civil Procedure Rules. The offer would remain for a specific time-limit and may be produced to the court in respect of costs.

There was the option of filing an application to Professional Arbitration on Court Terms (PACT). This would have had to be a joint application, and would only be relatively straightforward if all the terms of the lease except the rent has been agreed. The advantages of this would be flexibility, speed and cost.

The process of negotiations were mainly conducted via written communications. For negotiations to be concluded quickly, I should have arranged face to face meetings with the tenant's surveyor.

Reflective analysis

The project has enabled me to develop my competences of Valuation, Property Inspection, Drawings, Landlord and Tenant and Estate Management.

It has enhanced my own understanding of the legal process that governs lease renewals, including the service of statutory notices and the court process.

The property inspections have made me proficient in the measuring of retail units in accordance with the RICS Code of Measuring Practice.

As a result of the detailed analysis of the retail market, I have enhanced my knowledge of the rental values that are being achieved within the open market. In addition, I am able to devalue comparable evidence in terms of the type of transaction, location, size, level of demand, lease terms and the differences between primary and secondary locations.

I have gained experience by negotiating with another RICS surveyor and am able to conduct myself in a professional manner, with special reference being made to the latest edition of the RICS Rules of Conduct.

Appendices

The Appendices were: Appendix I – Photograph of Subject Property, Appendix II – Location Plan, Appendix III – Lease Plan, Appendix IV – Copy of Valuation, Appendix V – Comparable Evidence, Appendix VI – Photographs of Comparables, and Appendix VII – Copy of Lease. None of the Appendices are reproduced here.

Illustration of issues/possible interview questions

Q You referred to improving the investment value of the property through the lease renewal. How exactly would this be achieved?

Q What would you say were the key investment qualities of the property, assuming that the new lease was in place?

Q You made a number of 2.5% adjustments – where did you get the percentages from?

Q How would you describe the attractions and disadvantages of the property, and retailing potential, in the eyes of a prospective new tenant?

Q What is the potential for an A3 use, and did you investigate this?

Q Is VAT payable on the rent – if so, why/if not, why not?

Q Remind us what the key negotiating points were, and can you summarise how you deployed your negotiating skills (ie were there any particular tactics you adopted)?

Q You were fortunate that the main terms of the lease did not provide much scope for debate, but assuming there was a difference of opinion between the parties on some of the terms, what would the general principles be that the court and therefore the parties adopt?

Q Regarding the Civil Procedure Rules referred to, are there any pre-action protocols in place for lease renewal?

Q Also regarding the Civil Procedure Rules, what is the position regarding the award of costs, and how can this be different from the traditional position of the winner in the case typically being awarded costs?

Q Is there any particular case law you can refer to in support of this?

Q What if you as the landlord wanted a wider user clause in order that the rent could not be depressed with a restrictive user clause, and the tenant resisted this, how do you think the court might rule?

Q Incidentally, what are the rules on changing the lease area at lease renewal (ie the landlord excluding areas or adding to the demise)?

Q What steps did you take to ensure that the tenant was in 'business occupation'?

Q If you had required the property for redevelopment/refurbishment, what would your overall strategy have been?

Q You referred to PACT. Why is this not used very often?

Chapter 7

Tenant's Occupational and Business Issues

This chapter provides an illustration of the issues arising when a tenant needs to consider the best way of rationalising space, and minimising property overheads.

The commentary comprises an extract from *Estates Gazette*'s two-part feature of 6 and 13 September 2003, based on an APC event run by Birmingham Property Services, the in-house property consultancy of Birmingham City Council, in July 2003.

The landlord and tenant event was available to all general practice APC candidates. Prior to attending the event in Birmingham, candidates undertook a viewing of Alpha Tower, a 25-storey office in Birmingham city centre, part of which is occupied by Birmingham Property Services. An inspection was also undertaken of the immediate area.

At the event, a hypothetical scenario (unrelated to BPS's actual occupation) required delegates to assume that a tenant client is in occupation of floors two, three and five of Alpha Tower. The instruction advised that:

The client has suffered a downturn in business, and instructs the surveyor to advise on the options available to reduce property overheads, bearing in mind that only half the leased space is currently required. The client advises that a 15-year lease was granted with effect from July 2000, and there are two years to the first of the five-yearly rent reviews. Each floor comprises approximately 7,000 sq ft (650 m²) of modern/refurbished office space, with kitchen facilities on each floor, and toilets within the communal lifts area in the centre of the building. Other information will be for the surveyor to establish as part of the instruction.

Initial tasks

On receiving the approach from the client, it needs to be established that there are no conflicts of interest preventing the surveyor from acting. A conflict may arise, for example, if lettings, rent reviews, lease renewals, or other management work was already being undertaken on behalf of the landlord.

Terms of engagement will also need to be confirmed in writing with the client, to include the fee basis. In other cases, it may be necessary for the surveyor to check whether they have the expertise to undertake the instruction, and whether the task is within their professional indemnity insurance cover.

A copy of the lease will be required from the client, together with all relevant information details of separate agreements such as those in respect of parking, assignments and sublettings, consent for improvements, any works undertaken without consent, deeds of variation, previous disputes, service charges, and any correspondence/claims/legal notices from the landlord.

In addition to inspecting the client's demise, other parts of the property should be inspected where possible, and notes made of the surrounding area. Details will be taken of space to let within the building and in the immediate area, with asking rents if stated. A view should also be taken as to whether the property appears well managed, looking at staffing levels in reception, cleanliness, and the range and quality of facilities.

The client's business needs – finding the right property solution

It is important to establish the client's plans for its business, and understand the operational issues which influence its property decisions. This includes location, bearing in mind, for example, that less expensive edge-of-city rents may be an effective solution. Note, however, how some businesses rely on a city centre location it is accessible for clients.

Note also how an office move from the city centre to edge of centre can mean staff seeking employment elsewhere because of longer or more complicated journeys. For other staff, and also clients, the better availability of parking out of centre may be an attraction. Office specification, image, etc, will also have to be considered.

Relocation costs, possibly including double property overheads for a period of time, need to be considered, as well as the disruption caused while moving.

The client generally needs to be aware of the range of costs associated with vacant properties: rent, rates, service charges, internal repairs, insurance, security, etc, plus any surveyors', legal, marketing, stamp duty, etc, costs involved in property transactions. Property costs will represent a different proportion of total overheads for different client businesses, and may influence property decisions accordingly. In providing thorough advice to the client, consideration should be given to factors such as the scope to reduce rates payments by concentrating occupation into two floors and paying only empty/void rates charges on the other floor.

The client may also have other properties, either freehold or leasehold, which influence the options available. These include sales to realise capital receipts, and perhaps also lettings to secure income. It may be possible for the client to secure a loan against property assets (or increase an existing loan if property values have risen). Sales and leasebacks are another option, although a smaller, struggling, business is likely to be unattractive to the investment market, commanding a relatively high yield and therefore being an expensive option to the tenant. As well as the need to now pay rent, the tenant may not wish to enter such lease commitments. Some options will depend on whether the business difficulties are temporary, or represent a more permanent decline.

Accountancy and taxation issues may also affect property decisions, and it may be necessary to liaise with the client's accountant.

Discussions yield useful information

Discussions with the client's staff, other tenants and the managing agents may provide useful information as, sometimes, can talkative receptionists, security guards and caretakers. Occupier satisfaction, tenant turnover, location and so on will be important to gauge, and a list of occupiers from a display in reception and/or available space detailed on the agent's external 'to let' board may help to indicate vacancy rates.

The surveyor will need to assess the internal condition of the tenant's demise because although the landlord may not be pursuing remedies for disrepair, dilapidations may detrimentally influence any deal to relinquish space.

If any breach of the landlord's lease obligations can be identified, this may be helpful for subsequent negotiations.

Planning enquiries will also help to establish whether other schemes are taking place in the area, including those on-stream developments with planning permission secured.

In other situations, it would be more relevant to establish alternative use potential, such as A3 café/restaurant use for an A1/general retail shop, in order to make the leasehold interest as valuable and marketable as possible (lease terms permitting). In other cases, an alternative use, or even just scope for refurbishment, may help the tenant to secure a favourable deal with the landlord to vacate the property.

Also established will be information such as rateable value, and whether VAT is payable on the rent. Measurement will also take place. (Lease terms are outlined later.)

Tenants looking to take occupation of the property will make similar, usually more extensive, enquiries, perhaps including a building survey. Both parties will, of course, need to be up to date with the latest market transactions and space available.

Valuation and market issues

If not already known through active involvement in the local market, an indication of rental values will be gauged from the research on site, from enquiries with other agents and reported deals in the property press. A view will also be taken on the strength of the office market, and expected future trends.

In Birmingham, a contrast will be found with the falling rental values reported in the London office market over the past year or more. The development of Birmingham's city living residential market over recent years has squeezed the supply of office stock at a time when aggregate market demand for offices has been increasing. Prime rents, in the city centre at Colmore Row, touch £27.50 per sq ft, and at Brindleyplace, a standalone scheme about a quarter of a mile away, rents reach £24 per sq ft. Alpha Tower is situated midway, just outside the inner ring road which has traditionally been seen as a barrier between centrally located offices and those beyond, with market rents reflecting this.

Letting agent Lambert Smith Hampton reports that Alpha Tower is fully let, with the latest letting, for part of floor 2 in June 2003, achieving £16 per sq ft. In 2000, when the assumed tenant's lease commenced, quoting rents were £12 per sq ft up to floor 11 and £14 for floors 12 to 27; the difference represents the panoramic views of Birmingham.

Because rental levels for Alpha Tower have increased around 30% over the past three years since the assumed initial letting, the tenant client should secure a premium if assigning, or a profit rent if subletting. Profit rent/premium may also derive from the value of any improvements which the landlord cannot reflect in the rent.

If remaining in the property however, a relatively high increase in rental overheads seems likely at the next rent review in two years' time, thus having a detrimental effect on the client tenant's financial position.

It is worth noting that had a rent-free period been granted at lease commencement in 2000, the rent passing after the rent-free period (known as the headline rent) would be higher than the true day-one market rent stated above, and there would be less profit rent/premium.

In fact, an initial rent-free period could mean that the tenant's interest is currently overrented. If so, an assignment would involve paying a new tenant a reverse premium/capital sum, and a subletting would not be at a rent high enough to cover the rent payable to the landlord (if indeed a subletting were possible under such circumstances – see later).

Demand for assignment and subletting

In establishing the assignment premium, the surveyor needs to be aware that an assignment, with pre-determined rent and other terms (including a short, unexpired term), may not be as attractive to the market as a deal done directly with the landlord (where all terms are negotiable, rather than just the level of premium).

The demand for occupation by way of a subletting can be similarly unattractive in some cases: for example, if the headlease requires sub-tenancies to be contracted out of the security of tenure provisions of the Landlord and Tenant Act 1954, sub-tenants may be forced to vacate at lease expiry.

Such circumstances may, however, enable a prospective occupier's short-term needs to be met when landlords would insist on a longer lease for a direct letting.

Another factor that can make subletting relatively unattractive for prospective occupiers is the need to secure consent from both the immediate tenant landlord, and the head landlord, for works and other matters. Assignments and sublettings can also be subject to delay, with uncertainty as to whether consent will be granted, whereas direct deals with a landlord can be concluded relatively expeditiously.

With the rents quoted above, and as with comparable evidence generally, it is, of course, important to establish the lease terms, and especially the extent of any incentives (which would make the day-one market rent lower). Note also the effect of floor level on rental value as mentioned above, and the importance of surveyors picking up this point at letting, rent review, lease renewal, rating, etc.

Lease terms

In order to determine some of the options available to the tenant client, it is necessary to examine the lease.

Alienation provisions

Important early considerations include whether the three separate floors occupied by the tenant client are on the same lease, on separate leases for each floor, or on one lease for floors 2 and 3 and another for 5.

Leases often permit assignment of the whole only, so whereas three separate leases would provide optimum flexibility, assignment would not be feasible with a single lease of all three floors if the tenant wished to remain in occupation of part.

A single lease would also mean that if the tenant wished to relinquish its occupation by assignment, the accommodation could be marketed only as a whole 21,000 sq ft/1,950 m^2 on three floors, rather than as permutations of between 7,000, 14,000 and 21,000 sq ft/650, 1,300 and 1,950 m^2.

Subletting of whole or part is likely to be an option, but if subletting is not feasible, the tenant may have to vacate in order to be able to assign instead (noting also that subletting restrictions are sometimes more stringent than assignment provisions). Subletting of whole only would, of course, prevent partial occupation of the lease area. Such factors highlight the importance of obtaining flexibility, where possible, at lease commencement.

Although it may be possible to vacate and sublet all three floors (to a single tenant or several tenants), it may not be appropriate for the client to become effectively a landlord/investor, and carry the usual property risks, which may in turn create risks for the operational side of the business.

Break provisions

It should also be established whether the lease(s) contains break provisions for the tenant and, if so, when. A break at year five, negotiated at lease commencement, would be particularly advantageous. It would be important to establish whether there were any conditions relating to a break such as a penalty payment, or the requirement to be in compliance with all lease terms. Here, a break clause conditional on full compliance with repairing covenants can be a particular trap for the tenant.

Also, any break provisions for the landlord may influence the attractiveness, and therefore the terms, of any assignment or subletting.

User provisions

The user provisions are unlikely to present problems for offices (but may do for other uses, especially retail). Similarly, there are unlikely to be keep-open clauses which oblige the tenant to remain in occupation, and are more common with retail properties. It is worth checking also that there is not a surrender-back provision, obliging the tenant to offer the landlord a lease surrender before seeking to assign.

Longer-term liabilities: privity of contract

The client will need to be advised of any ongoing liability in the event that the lease(s) were assigned. Because the lease commenced after January 1996, ongoing liability through privity of contract will not present a risk, but if there is a requirement for an authorised guarantee agreement in the lease, following assignment, the current tenant will be liable for rent and the performance of other covenants in the event that the new tenant defaults. The new tenant therefore needs to be selected carefully. 'Mainly for Students' 26 January 2002, p164, provides more detail on this, including the effect of the Landlord and Tenant

(Covenants) Act 1995, Landlord and Tenant Act 1988, and other aspects relating to assignment and subletting.

Rent review provisions

As an example of the importance of scrutinising all lease terms, it would be necessary to check that the valuation basis under the rent review provisions was to open market rental value, and not for any reason, albeit unlikely in the case of city centre offices, on an alternative basis. (Clauses requiring the review of a hypothetical building can cause particular problems.)

Surrender and other options

In addition to assignment and subletting options, possibilities for the tenant include the following.

Negotiated surrender

Landlords will not often be willing to accept a surrender, involving loss of rent and having to cover the other costs of vacant property, as well as the uncertainly as to when, and on what terms, a new tenant can be secured. This would also, of course, affect the capital/investment value of the landlord's interest (albeit perhaps temporarily until a new tenant is found). This would be particularly important if the landlord was planning to sell the property, or refinance against its capital value (including to realise equity for other purposes, such as further acquisitions).

However, such may be a tenant's anxiety to relinquish its leasehold interest that the landlord may still take the opportunity to secure a surrender on competitive terms.

The tenant's wish to relinquish its interest may be particularly beneficial for a landlord of multi-let offices where the opportunity can be taken to refurbish dated specification, and achieve a higher rent and more favourable other lease terms, also boosting capital/investment value.

Facilitating a new lease

Although the landlord may not wish to agree a surrender, for the reasons mentioned above, it may benefit both parties if the tenant is able to market the property on the basis that a new lease may be available to a new tenant.

The tenant client's interest may then be relinquished more easily, and the landlord has the chance to agree terms that have a more beneficial effect on investment value.

The tenant may still receive a surrender premium from the landlord or have to pay a surrender premium to the landlord, with the new lease being granted contemporaneously with the surrender of the old lease. This would also ensure that the tenant's liability, through privity of contract, or an authorised guarantee agreement, ends (unless the landlord requires the tenant to act as a guarantor within the lease to the new tenant).

Landlord markets the property

Another option is that the landlord markets the property himself. This is not common (especially in multi-let situations where the landlord may have his own voids to market), but may be beneficial, for example, in the case of single-let properties with tenants of poor covenant strength whose imminent insolvency/departure would end the receipt of income, with little chance of debt recovery through the courts.

Other possibilities

These include the tenant surrendering the existing space, and taking a new lease of a smaller area either in current lease area, elsewhere in the building, or even in another of the landlord's properties. Other tenants in the building could be approached to see if there were any mutually beneficial possibilities such as another tenant with an imminent lease renewal taking a subletting from the tenant client rather than a new lease from the landlord. An assignment could be undertaken similarly. Although it may technically be possible to assign, and sublet part back, this is not common.

In other situations, it may be possible to buy the freehold, by negotiation or sometimes by exercising an option to purchase within the lease. An option to renew may also influence negotiations and marketability of the leasehold interest.

Marketing

The actual marketing of the tenant's interest in a particular building is outside the scope of this feature, but another difficulty associated with relinquishing space on behalf of the tenant in multi-let property is that if there is other space available (either direct from the landlord, or other tenants seeking to assign/sublet), those parties introduced on behalf of the tenant client may see other opportunities on viewing, obtain details and pursue negotiations elsewhere.

Another check that the surveyor would have to make is whether site boards are permitted under planning regulations (this is necessary because the property may be listed or in a conservation area), and under the terms of the lease.

Assignment and subletting – investment and management issues for the landlord

The landlord will be keen to ensure that the security of his investment is protected, and that a new tenant is of good financial standing/covenant strength. In the case of subletting, the landlord will, of course, retain the current tenant, but will be keen to ensure that subtenants are appropriate occupiers for the building. Consider the impression created, for example, in a building filled with suited professionals, by a tenant whose less-attired staff clutter the entrance areas on cigarette breaks. Lease terms may not prevent this.

The landlord will also be anxious to ensure that the terms of sublettings do not have a wider detrimental effect on his interest. Examples include the terms of a subletting being used as comparable evidence against the landlord for rent reviews and lease renewals in the building, and rents agreed on subletting similarly influencing capital valuations, and therefore potential sale price and/or the ability to raise finance.

It is important that the surveyor acting for the tenant client evaluates the landlord's ability to refuse consent, thus avoiding provisionally agreeing terms for deals that the landlord is unlikely to accept.

Assignment and subletting provisions

At the event and within the *Estates Gazette* article, reference was made to a previous *Estates Gazette* article on assignment and sub-letting, which is now included in *EG Books'* publication, *Best of Mainly for Students*, vol 3. This includes further detail on assignment and sub-letting lease provisions.

Keeping up to date with case law – difficulties with subletting

At the BPS event, guest speaker Anne Waltham of Wragge & Co LLP, Birmingham, outlined the implications of *Allied Dunbar Assurance plc* v *Homebase Ltd* [2002] EWCA C iv 666 on subletting. This was a good example of how general practice surveyors draw on case law in their day-to-day work.

The decision

The Court of Appeal refused a tenant permission to use a supplemental deed (which was expressed to be personal to the parties), to circumvent the restrictions on underletting contained in the tenant's lease.

The facts

The tenant had a 25-year lease of a large retail warehouse which precluded underletting, unless:

- At full market rent without taking a premium.
- On substantially the same terms as those of the lease.
- With rent review provisions to mirror those of the lease (unless the underlease was to expire before the next review under the lease).
- The landlord consented to the proposed transaction (such consent not to be unreasonably withheld).

The tenant found it difficult to find a suitable subtenant. The eventual candidate was not prepared to pay a rent as high, or accept repairing obligations as onerous, as those contained in the lease. This was reflected in the heads of terms. The form of underlease satisfied all the pre-conditions laid down in the lease, but was subject to the provisions of a collateral deed, expressed to be personal to the tenant and its undertenant, pursuant to which the tenant undertook to pay the difference between the rent that was payable under the heads of terms and the rent reserved by the underlease; and the cost of complying with certain repair obligations. The landlord refused consent to the subletting and the undertenant declined to proceed with the transaction. The tenant sought damages for breach of statutory duty under the Landlord and Tenant Act 1988.

The Court of Appeal judgment

The Court of Appeal found in favour of the landlord. The court confirmed that the collateral deed and underlease were interdependent and had to be read together. It made no difference that the deed was expressed to be personal to the tenant and the undertenant. The original parties were to be taken to have intended that the landlord should be able to control the terms of any permitted underlease. The terms of the underlease might become the terms of a tenancy between the landlord and the undertenant, on the termination of the headlease. Moreover, the rent under the underlease was an obvious comparable with any rent review under the headlease.

Implications and problems

- This is a landmark case for landlords and tenants thousands of leases contain a similar prohibition on underletting at less than the passing rent.
- Would the result have been different if the tenant and undertenant had contracted out of the 1954 Act? Landlords would argue not.
- The decision makes it doubly important for tenants to review their position carefully at an early stage, especially if they have the option of exercising a break clause in circumstances where market conditions are weak.
- In the rent review context it must be strongly arguable that a prohibition on underletting at less than the passing rent will be considered onerous at rent review and will merit a discount.

Keeping up to date with case law: landlord's obligation to act reasonably and not delay

At a separate event in January 2004 – a Landlord and Tenant update run by RICS West Midlands CPD Foundation and Advantage West Midlands – Anne Waltham of Wragge & Co LLP outlined the case of *Blockbuster Entertainment Ltd* v *Barnsdale Properties Ltd* [2003] EWHC 2912 Ch. This highlighted the importance of landlords dealing speedily with tenants' applications for consent. and was subsequently covered by *Estates Gazette* as follows.

Blockbuster was the tenant of premises in Canterbury, held under a lease with over 10 years to run. An application to the landlord for consent to underlet the upper floors to a local college was made on 28 May 2002. The landlord consented to the grant of the underlease in principle on 15 July after strong pressure from the tenant, by which time the college had already withdrawn from the transaction.

The court considered whether the landlord was guilty of an unreasonable delay in dealing with the application. The landlord had asked the tenant for a certificate in the form required by the lease, and for references for the college. The court ruled that the landlord could not have been accused of being dilatory if it had done so 'on receipt of the tenant's application', which was made on 28 May, 'or within, say, a week of that date'. But it did not do so until two weeks had gone by (on 10 June), by which time it had instructed solicitors. The tenant dealt with the landlord's request by 19 June, when the landlord had all the information asked for and that was relevant.

The High Court took the view that the landlord should have given its consent, in principle, within the week, that is, by 26 June. The judge went on to rule that the landlord could also have been expected to consent in principle to a change of use and to the college's plans to alter the premises by the end of June or at the beginning of July. He said: 'In those circumstances, even if the documents had not yet been executed, they would have been ready for execution...The College would in that event not have withdrawn from the transaction...pending completion of the under-lease.'

The landlord was liable for damages for lost rent from 1 August 2002 to 4 August 2003 (when the tenant did finally underlet the premises), for rates for the upper floors for the same period, and for a contribution to the insurance of the building, totalling £72,000. Fatal errors by the landlord included:

- The landlord's agent telephoning the tenant to ask it to surrender the lease in order to facilitate conversion to residential use.
- The landlord taking the view that it had no statutory duty to deal with the tenant's application because it was not accompanied by a certificate in the form required by the lease. The court refused to accept that this was a condition precedent of the lease. The landlord could legitimately ask for a certificate if one was not supplied, but had a duty to deal with the tenant's application in the meantime, and was under a duty to ask for the certificate with reasonable speed.

Case law for the APC

As mentioned in the article covering BPS's event in respect of Alpha Tower, the criteria for the knowledge required in respect of case law is major and current case law, case law referred to in written submissions, and case law supporting key facets within the critical analysis.

Chapter 8

Landlord and Tenant, Surrender, Improvements, Public Sector Issues

In this chapter, an example is provided of a surrender of a lease which also considers other issues, including tenants improvements. The case also shows the work in which a local authority may be involved, and how a candidate in the public sector can demonstrate commercial awareness, as well as covering the various technical issues in due depth. The report is contributed by Kitt Walker.

Introduction

This critical analysis examines the negotiation of a surrender and regrant of a lease of nursery accommodation, undertaken while working for my previous employer.

The case has been selected for its interesting combination of property, business and political issues, and because it allows me to demonstrate my skills in competency areas of landlord and tenant and valuation in particular.

I can confirm that the consent of my previous employer and the incoming tenant have been obtained to the disclosure of information within this report. The outgoing tenant was not approached owing to separately continuing negations with my previous employer, and their identity has consequently been withheld.

My report will initially outline the nature of the instruction, and examine the case objectives and key issues. It will then cover the principal factors affecting the surrender settlement and the deal with the new tenant, and will conclude with a reflective analysis of the experience gained.

The instruction

My client in this case was the Education Department of the county council. The terms of instruction and fee basis fell within an existing agreement covering all consultancy services provided to the client.

The nursery accommodation was located in a small village in the county, and principally served the adjoining school. The accommodation comprised a temporary classroom with a gross internal area of 60 m² and use of the play area, car park and toilets.

The current tenant had occupied the property on a five-year FRI lease from January 1999. The initial rent of £1,750 was subject to an upwards-only rent review, due on an RPI basis after three years. The lease was contracted out of sections 24–28 of the Landlord and Tenant Act 1954.

As part of the initial letting, the tenant had relocated the nursery accommodation, which was owned by the council, from another site, and also incurred substantial improvement expenditure in order to make it useable.

In late 2000, the current tenant no longer wished to operate from the site owing to more favourable business opportunities being considered to be available elsewhere. The lease prohibited assignment and subletting, which together with a tightly drafted user clause and the absence of a break clause, meant that it would not be possible for the tenant to relinquish his interest without my client Education Department's agreement.

It was important for the client that nursery accommodation was available to support the adjoining school, and the local area more generally. This relieved any pressure to provide such services, and while nursery provision would not have been obligatory for the council, the presence of a private operator helped avoid considerable facilitating costs, and any issues of subsidy.

Case objectives and key issues

It was my responsibility to agree terms for surrender with the current tenant, and secure a new letting with a new tenant. The new lease terms had to be on an open market basis, adhering to my client's legal requirements in respect of obtaining best value.

The key issues were:

- First, the need to establish the level of compensation payable to the outgoing tenant. This would reflect any profit rent which may have accrued in market value terms since the commencement of the lease.
- Second, and more significantly, compensation would reflect the value of the improvements undertaken by the tenant.
- Third, the acceptance of a surrender from the outgoing tenant had to be conditional on a new tenant paying the compensation by way of a premium (and ensuring, overall, that market based values for the premises were upheld, whether by rental, premium or ownership of improvements, or a combination of these).

Another important aspect of the case was to regularly report back to my client department, and also keep the head teacher and governors of the adjoining school informed of progress.

Establishing the nature of the tenant's interest and factors affecting compensation

In order to establish the amount of compensation that my client may be able to pay on an open market basis (as opposed to any bullish ransom basis promoted by the outgoing tenant), I had to assess value on both a 'profit rent' basis, and on a 'compensation for improvements' basis.

The rent passing, set in January 1999, was £1,750. I had to investigate the background to the transaction in order to be satisfied that this represented a true market rent, and establish the assumed condition of the property (ie including or excluding the tenant's improvements). Account had to be taken of any other possible elements of the deal, such as any payment to relocate the building and undertake improvements. An assessment also, of course, had to be made of the current day market rent.

I first established that the improvement works had not comprised an obligation in the lease. This would have defeated any entitlement to compensation for improvements. Although there was not, strictly, any entitlement to improvements compensation at mid-lease surrender, I was aware that such compensation would indeed be available to the tenant at the expiry of the lease, and therefore affect the value of his current interest.

It transpired that the tenant had negotiated strongly regarding the exclusion of a works obligation clause at lease commencement, and my former colleagues had been unable to secure such an obligation. It was apparent that the outgoing tenant exploited a strong negotiating edge at the commencement of the lease – reflecting my client's requirement to secure nursery provision from a private operator. It was a this point that I became wary that the outgoing tenant may undertake a similar approach as part of surrender negotiations, especially as he had already indicated a willingness to vacant, and continue paying rent if need be.

It was also established that the rent for the accommodation was set having regard to the unimproved condition of the property. It was accepted that the absence of any works obligation was fair to the tenant.

It is also worth mentioning that I had to check whether compensation for improvements would be payable at lease expiry, even though the lease was contracted out of sections 24–28 of the Landlord and Tenant Act 1954. Improvements compensation would indeed have been payable, although obviously any rateable value based compensation for disturbance would not be, owing to the contracted out status.

In establishing the case background, I also discovered that:

- My client department had paid the tenant to relocate the building from an alternative site to the subject site. This did not influence the surrender settlement being negotiated, as the cost was small. Also, a nominal rent-free period covered the time involved in setting up the property, and again did not influence the subject surrender.
- The initial cost of improvements were £19,948.05, say, £20,000.

Valuation

In view of the marginal increase, if any, in the market value of the property since the initial letting, and the absence of direct comparable evidence, the tenant's leasehold interest was not considered to have any value on a profit rent basis, other than the element attributable to improvements.

I then identified that in respect of improvements, the value of the tenant's leasehold interest on a profit rent basis could comprise a capitalised profit rent only for the duration of the lease. This was because while contracting out did not prevent compensation being payable for improvements, it would have prevented the tenant obtaining a further lease at lease expiry, thus enabling the landlord to secure a full market rent, including for the improvements (or alternatively, of course, requesting that the tenant vacated). This was in contrast to the more usual position whereby the tenant would benefit from the value of improvements beyond lease expiry, owing to the security of tenure afforded by the Landlord and Tenant Act 1954, and the effect of the 21-year rule.

However, in addition to value of the profit rent deriving from improvements until lease expiry, the value of the tenant's leasehold interest had to reflect the entitlement to improvements compensation at lease expiry. Such compensation would, however, have to be discounted to present value, otherwise the position would not be realistic (ie compensation paid now could be invested by the outgoing tenant to earn a return, hence the need for discounting to present value).

The amount of compensation payable, typically at lease expiry, is set out under section 1 of the Landlord and Tenant Act 1927, and provides that compensation shall be the lower of:

- The net addition to the value of the holding as a direct result of the improvements.
- The reasonable cost of carrying out the improvements today, subject to deduction for any wants of repair.

Owing to the limited and variable comparable evidence, it was difficult to meaningfully establish the difference between the unimproved and improved rental value, and appropriate yield and multiplier. It was at this stage thought reasonable to assume that the cost of the improvement works would equate to the increase in value they added to the property in order to bring the property in to use.

It was therefore necessary to establish exactly what the cost of improvements would be today. By liasing with my colleagues in the building section, it was felt that an appropriate sum would be £22,000. This was slightly higher than the tenant's initial expenditure of £20,000, but reflected the fact that rates would have increased with inflation, and that procuring the works in the market would be more costly than the amount incurred by the tenant.

This figure of £22,000 appeared to be too high an amount to pay for compensation, so I reconsidered my initial view that the cost of the works may equate to the increase in the value of the property. I had to distinguish between any amount which was purely property based, and any additional amount which the tenant may have felt beneficial to spend from a business perspective. This was because compensation for improvements is based on property values, and does not reflect business income and profitability. (It could have been argued that a profits method could be applied to nurseries, but the relatively uniform accommodation, and rent per sq ft basis of value, albeit with limited comparable evidence, meant that it was not appropriate.)

What I did identify was that the addition to the value of the property would not have been similar to the cost of undertaking the improvements, because the building was nearing the end of its economic life, and would provide only 10 years, or slightly more, of continued occupation. So while my initial approach was correct in principle, the short life span of the building rendered it inappropriate.

It did in fact seem that the tenant may have unwisely spent £20,000 at the outset, not fully recognising the life expiry of the building, and/or misjudging the profitability of the business. This was borne out by the tenant's stated reason of requiring a surrender – that there were better business prospects elsewhere. These findings presented me with a negotiating strategy upon which I would in due course rely – ie that some of the initial expenditure had represented the tenant's view of business prospects, and were unrelated to the increase in value of the actual physical property. In order to avoid unsettling relationships, I did, of course, have to avoid suggesting that a poor business decision may have initially been made.

Other factors accounted for in the valuation

It was noted that by the time that lease expiry would be reached in over three years' time, that either an increase in the cost of undertaking the improvements, or an increase in the value that improvements added to the property, may increase the amount of compensation due. It was not considered appropriate to reflect this fact by uplifting the compensation figure above today's value.

Also, a suitably low yield would be applied in respect of discounting the value of improvements at the end of the lease to today's value. This was considered to be in order, as it was known that the tenant, and also typical tenants, would been in a positive cash position, with the opportunity cost of capital effectively being bank or building society rates. Furthermore, such properties did not typically command an investment yield, but if they did, would be in the region of 15% or more owing to the special nature of the property, lack of alternative tenants, and risk of similar ventures in the area pulling away custom. In other words, an investment yield did not equate to the characteristics of occupational demand.

Another factor influencing the want for a straightforward approach to account for the possibility of improvements compensation rising over the period to lease expiry, was that there was, in fact, an RPI rent increase due after three years, ie in 18 months' time. This would have been small owing to modest levels of inflation.

This off-setting of potentially complicated elements had proved effective in rent review and lease renewal cases in which I had been previously involved, albeit obviously involving different variables. It ensures that the valuation approach does become disproportionately precise when valuation ranges, and also the range in which a negotiated settlement may lie, are considerably wider.

Calculating the compensation and negotiations

I reported to the client that the tenant had requested a sum of £20,000, representing the full extent of improvements expenditure incurred on the property. Many of the above points were also outlined to the client in respect of the legal entitlement to compensation for improvements, together with acceptance of the commercial reality of the situation – namely, that the client would have to pay a certain amount in order to secure possession, secure a new tenant and guarantee the continued provision of nursery accommodation.

It was also outlined to the client that while compensation in terms of property interests and the relevant law produced a low figure, account also had to be taken of any loss in value to the tenant's business/profitability. This was not however thought to be significant in this case, otherwise the tenant would arguably be wishing to remain in business.

The outgoing tenant maintained arguments in respect of improvements expenditure having been incurred, and due recompense being required. It was pointed out that the tenant was free to remain in the property, and that the client simply could not afford to pay an amount above £5,000.

I found it fascinating to examine the legal and valuation aspects in detail, but then be faced with a 'horse deal'. However, if a deal could be secured with an incoming tenant, that would be strong evidence of market value for the existing interest.

In order to meet the client's objectives of avoiding having to effectively subsidise the continued operation of the nursery, either by grant funds or sub-market rents, a deal with an incoming tenant had to be contemporaneous with the surrender, and on acceptable market terms. I have not outlined the deal with the incoming tenant, except to report that a premium of £5,600 was secured (which covered the compensation required to the outgoing tenant). In return, a rent-free period of three years was granted, within a five-year lease, with a rent review taking place after three years to £1,750 pa (the rent with the outgoing tenant), plus RPI.

I managed to negotiate a compensation figure of £5,000 with the outgoing tenant. Although, the overall deal produced a deficit of three years' income at £1,750 pa, it avoided the client having to find new funds, and vested the value of the improvements in the client's freehold interest. At the end of the five year term, the rent would be based on the improved condition. Although, this could have been commanded from the outgoing tenant owing to the contracted out status of the lease, and the 21-year rule not applying, it would not have been commercially, or politically, feasible to take such a course of action with the tenant. The new tenant was aware of the longer term liabilities from the outset, enabling a higher rent to be secured in due course.

I also managed to include a 'keep open' clause in the new lease in order that the incoming tenant would continue to operate the nursery, and not cause the difficulties encountered with the outgoing tenant. A keep open clause had not been included in the previous tenant's lease.

The surrender was undertaken by 'operation of law', with the original lease and keys being handed over to me by the outgoing tenant. Confirmation was provided to the client department and their solicitor. The payment from incoming tenant was made directly to the outgoing tenant.

Other options available

In addition to establishing the possible terms of a surrender settlement, I evaluated alternative options, relating to the outgoing tenant's possible breach of covenant.

Dilapidations, and other covenants

It was necessary to assess the current condition of the property, and determine whether the outgoing tenant was liable for any disrepair. Aside from perhaps making forfeiture proceedings viable, this would have helped reduce the compensation payable, noting also that the incoming tenant would have required due allowance to remedy any disrepair. The property inspection found the accommodation to be well maintained, and the tenant to be in compliance with all other lease terms.

Forfeiture option

I was aware that a means of securing possession would have been to physically repossess the property had the tenant been in rent arrears, and there was a forfeiture clause in the lease. From my management work, I had been aware that the tenant was once previously in arrears. A check of the rent payment history, and current arrears indicated that the tenant was late with the 1 October to 31 December quarterly payment, and forfeiture was in fact a possibility. (Rent was due in accordance with the modern quarter days.)

I discussed this option with my manager, and also the client department, but it was considered inappropriate from a political perspective to repossess the nursery. This was despite the tenant having just ceased to operate the nursery, and the school in particular being anxious for a new operator to be found, and the nursery to re-open. I was also aware of tenants' rights to claim relief against forfeiture.

Interest on arrears

The only negotiating edge I was able to find in respect of rent arrears was that interest of 4% pa above the Lloyds Bank plc base rate would have been payable on arrears. While the council do not invoke such a penalty automatically, I have found that its mention helps concentrate the mind of tenants, and payment is usually forthcoming (except in cases where payment is genuinely unaffordable).

Reflective analysis

In addition to the analysis provided in the main text above, the case provided good experience in respect of landlord and tenant issues, and rental valuation in particular – including the calculation of improvements compensation.

The biggest thing I learnt from the case was how on the detailed consideration of the various issues, including valuation analysis, more negotiating factors become apparent than originally appeared to be the case. This demonstrated the importance of always thinking through case options, and their respective detail, from the outset.

In hindsight, I should have investigated all of the facts before discussing possible terms with the outgoing tenant. Although I was only avoiding talking about figures at the time, I felt that had I been able to justify why I disagreed with the outgoing tenant's proposal, rather than say that I would look at the position, the outgoing tenant's hopes for a high compensation settlement would not have been raised. This did not however affect the ultimate settlement, which I consider to be successful, and a reward for the hard work applied to the case.

It would have been helpful if further comparable evidence was available, but I do not feel that anymore could be done to secure it. The client already had evidence from within their portfolio of similar accommodation, and I was aware of other transactions undertaken by private landlords.

It could be argued that the profits method of valuation and analysis may have been more appropriate than a straight comparable approach, but I still consider that the right approach was undertaken. Reasons for this included the limited range of comparable evidence and the absence in the market of a recognised percentage to apply to turnover/profits, leases not containing the right to obtain the tenant's accounts and therefore establish turnover/profits, profitability typically being dependent of the calibre of the operator rather than the property itself, and the difficulty in making adjustments such as to reflect the start up nature of some nurseries.

The negotiating techniques and awareness of business issues have more recently served me well as part of the work undertaken at Advantage West Midlands, my new employer. This includes an understanding of how the property issues, and respective values, may be only part of what a client is trying to achieve more widely.

Appendices

The Appendices were: Appendix A – Site Plan of Primary School, Appendix B – Photographs, Appendix C – Summary of Leases, and Appendix D – Comparable Evidence. None of the appendices are reproduced here.

Illustration of issues/possible interview questions

Q Had you been working in private practice, what might you have done differently regarding the initial terms of instruction?

Q You referred to the Education Department as being an internal client. How is the council structured in terms of the property team with whom you work and the various departments – and what are the reporting and approvals procedures which apply?

Q What use class does a nursery fall under, and are there any other uses to which the accommodation could be put within the same use class?

Q Can you describe the property, and in particular, was it really a temporary building?

Q What is the planning situation regarding temporary building?

Q You referred to the lease being contracted out of sections 24–28 of the Landlord and Tenant Act 1954. What are those sections exactly?

Q Why was the rent review basis RPI?

Q Why was assignment and subletting prohibited?

Q Regarding compensation payable to a tenant – how is that treated for tax (income tax for an individual or corporation tax for a limited company)?

Q You referred to a 'keep open' clause. I recall case law on this involving a supermarket. What was this, and what is the position in the event that the tenant ends trading (ie can the landlord obtain an injunction to force the tenant to remain trading, or would damages be due to the landlord)?

Q You referred to the 21-year rule. Can you explain again why was this not relevant – ie if the tenant took on a new lease at lease expiry, the value of improvements would not be reflected?

Q How would you advise a tenant client taking on a lease which the landlord required to be contracted out – ie what are the pitfalls?

Q Where did the new tenant come from – were you also marketing the property?

Chapter 9

Retail Assignment on Behalf of Landlord Investor

This chapter considers an assignment of retail premises on behalf of a landlord/property investor. In dealing with the tenant's application to assign, the landlord will be particularly conscious of the financial standing of the new tenant, and the impact that this could have on the capital value of the property. The report is contributed by Ruth Walsh.

Introduction

My critical analysis focuses on the assignment of retail premises with which I was actively involved on behalf of a client landlord.

I have chosen this assignment as I feel it demonstrates my ability to carry through a project and apply the technical and practical knowledge that I have gained during my training period.

After providing a brief background to the project, this report describes the processes with which I was involved, identification of client objectives and key issues that needed to be addressed in order to consider the assignment. Finally, there is a critical appraisal of the outcome and a reflective analysis on the lessons learned and experience gained.

In accordance with RICS Rules of Conduct, I confirm that I have obtained the consent of my client, the assignor and the assignee, in order to disclose matters that I have outlined within this report.

Background

The retail property that is the subject of the assignment forms part of a large portfolio of mixed property types owned by my client, and which include office, retail and industrial interests. The client is a large pension fund for whom my firm is retained as managing agents, and who are involved in the day-to-day management issues of the fund on behalf of the client.

Instruction

I received a letter from the assignor's agent dated 10 April 2003, seeking consent to assign the whole of the premises to a new tenant. The assignor's agent applied in writing to my firm directly, as we act on behalf of the client (the assignor's landlord). The assignor was Retail Sport UK. They had recently been acquired by the Sports Group plc, and were seeking to assign to Soccerstores Ltd.

Conflict of interest

Prior to accepting the instruction, it was necessary to check on any potential conflict of interest. In compliance with my firm's ISO 9001 Quality Assurance Procedures it was important to establish whether any potential conflicts of interest existed. I completed a conflict of interest check through my firm's electronic database. No conflicts were found.

Agreement of fees

The assignor's solicitor had requested an estimate of my client's surveyor's and legal fees in their letter seeking assignment. I confirmed in writing to the assignor's solicitor the respective fees, including VAT, which I proposed.

The assignor's solicitor undertook to meet the surveyor's fees plus reasonable legal fees incurred in the transaction, whether or not the assignment proceeded to completion.

The surveyor's fees would be that of my firm and the quotation was based on the standard assignment fee. The solicitor's fees related to my client's solicitors and I quoted for reasonable legal fees to be paid.

I was reluctant to agree a cap on my client's legal fees as this may prejudice my client, eg Soccerstores Ltd could argue points of the assignment document.

Under section 1 (3) Landlord and Tenant Act 1988, my client had to give a decision in writing within a reasonable time frame as to whether consent to the assignment would be granted or grounds for refusal given.

Property description

The property comprises mid-terraced, single-let, retail premises of mid 19th-century construction. It is located on High Street, Guilford, which is a pedestrianised cobbled street forming the heart of the prime shopping area of the town.

The property is arranged over ground, first and second floors with the second floor being used as a storage area. It extends to some 751 m^2 (8,084 sq ft) of total floorspace with sales floorspace equating to 641 m^2 (6,900 sq ft) and the remainder being ancillary floorspace of 109 m^2 (1,173 sq ft).

A Location Plan, Street Plan, Ordnance Survey Extract and Photographs are attached at Appendix I, II, III & IV respectively (not reproduced here).

Client's objectives

My client's main objective was to maintain the investment value of their interest in the premises.

In order to achieve this objective, the client needed to understand the character and financial standing of the proposed assignee. The client also needed to preserve the security of rental income by controlling the identity, use and type of potential assignee.

The covenant strength of the proposed assignee would need to be the same or stronger for the client to accept the assignment.

The client was obliged under the Landlord and Tenant Act 1988 to:

- Give consent to an assignment, except where it is reasonable not to do so.
- Give written consent without undue delay.
- Demonstrating reasonableness if refusing to grant consent or imposing any conditions to the granting of consent.

Key issues

There were a number of initial key issues that needed to be considered, including:

- Check the alienation provisions in the lease to establish whether the lease provided that the tenant could assign their interest.
- Establish whether the alienation provisions were absolute, qualified or fully qualified.
- Check the user clause covenant within the lease to establish if it was restrictive or was open within the use class.

Once these initial key issues had been considered, I noted further issues and information, which would be required to enable both the client and myself to consider the application. These were:

- To review the assignment clause within the lease to determine whether conditions are attached to consent of assignment.
- To determine if the lease specifies circumstances where the landlord can withhold consent to the assignment.
- Whether or not an Authorised Guarantee Agreement (AGA) would be required as the lease was defined as a new lease under the Landlord and Tenant (Covenants) Act 1995.

Having considered the proposed assignment, I was of the view that the following was required:

- Three years' audited company accounts for the proposed assignee.
- Details of parent companies and sureties of the assignee. (The proposed guarantor.)
- Heads of Terms between the parties, premium or reverse premium payable, and confirmation that there is no side letter or deed proposed.
- Details of any alterations the proposed assignee would wish to carry out.
- Details of other units where the proposed assignee trades, photographs of these units (question: is this reasonable? Maybe not.) and the company's business plan.
- Details of any premium agreed between the parties.

Lease analysis

My first step in the process was to check the lease, to determine what the tenant was allowed to do by way of assignment and to check the user clause provision. I checked the conditions that the assignor and proposed assignee would need to observe as part of the consent to assignment. I also checked the circumstances under which consent could be withheld.

The checks I carried out are detailed in the following paragraphs.

Alienation

The alienation provision included in the lease provided that the tenant was not to assign any part of the premises. Instead the lease contained a fully qualified covenant in favour of assignment of the whole of the premises: 'Not to assign the whole of the premises, without prior written consent of the landlord, such consent not to be unreasonably withheld'.

User

The user clause governs the use to which the tenant can put the property. An application for change of use was not required. The classification within the use class order was for Class A1 use. The proposed assignee was a clothes retailer and, as such, restrictive user provisions would not affect the assignment.

Conditions attached to consent

As with many modern leases, the assignor was required to agree to an Authorised Guarantee Agreement, to guarantee the performance of the proposed assignee.

At this stage I consulted The Code of Practice for Commercial Leases in England and Wales (2nd ed, April 2002) produced by the Commercial Leases Working Group 2002, which contains a recommendation (Recommendation 9) that landlords, when being asked to agree to an assignment, consider requiring AGAs only where the assignee is of lower financial standing than the assignor at the date of assignment.

The assignor had to pay any arrears of rent and other sums due to the landlord prior to the date of assignment. I ran an arrears schedule on Advanced Program Base (APB) Manager (this is a property management database, which allows the surveyor to call up information relating to a tenancy, such as lease details and rent and service charge information). The assignor had no arrears outstanding.

My client was entitled to require the assignee to have a surety or sureties that were acceptable, and would covenant to guarantee the assignees covenants under the lease.

Circumstances to withhold consent

The lease identified the following circumstances where consent could be withheld. These were:

- If my client was of the opinion that the proposed assignee was not of sufficient financial standing to enable the covenants in the lease to be complied with.

- If the proposed assignee was able to claim diplomatic immunity.

The burden of proving that consent should be withheld was on my client.

Summary

My analysis of the lease provided the primary source of information required for the assignment. I had now ascertained the conditions which the proposed assignee and assignor would have to meet.

The user clause was unrestrictive within Class A1, and only the whole of the premises (as opposed to any part) could be assigned.

In order for me to consider the assignor's application, I requested that the assignor supply the items of information listed in the second tranche of key issues. At this stage the assignor and assignee had agreed a time scale for completion of the assignment of 2 June 2003.

Accounts

The information I had sought on the financial standing of the proposed assignee came in the form of three years' audited financial accounts on Soccerstores Ltd. In addition, one year's full audited accounts was provided for Soccerstores International of Germany, who agreed to act as guarantor for Soccerstores Ltd.

I had requested accounts of the proposed assignee's guarantor to be provided as there were 12 years left on the residual of the term. The tight time scale given prompted me to request all information.

Covenant strength

I ran a Dun & Bradstreet report on the assignor and assignee. Dun & Bradstreet is a provider of business information and provides detailed financial information on companies' financial strength by analysing turnover, profit/loss, and net worth. An explanation of D&B terms is attached at Appendix V (not reproduced here).

The Sports Group plc had a rating of 5A1. This indicated that the Group had a financial strength of £35+ million (based on tangible net worth) and represented minimal risk. It trades from over 150 stores throughout the UK and is listed on the FTSE SmallCap Index. Turnover to 31 Jan 2003 was £371 million with pre-tax profit of £10.8 million.

Soccerstores Ltd had a D&B rating of A2. Its financial strength was £350k – £700k, and it represented low risk. Soccerstores Ltd's guarantors (Soccerstores International) represented minimal risk. The company's annual report detailed the company's future prospects, and it was also established that Soccerstores had expanded its retail format globally. It plans to open 62 stores in Europe in 2004, 25 of which will be located in the UK. The company currently has 361 retail stores in Europe.

Analysis of accounts

The accounts were reviewed, noting in particular, net profits after tax from the profit and loss sheet, and current assets on the balance sheet for each year in question. In establishing the financial standing, certain items were considered, including profitability, liquidity and gearing. Trend analysis of various ratios were considered, including the current ratio, acid test ratio and debtors days. Year-to-year comparisons were also made.

I reviewed the profit and loss accounts and the balance sheet to ascertain the assignee's current financial strength with the assistance of a Senior Funds Accountant within my firm. Notes to the accounts, including the director's report and the chairman's statement, provided useful information about activities in which the company had been involved. The net assets and profitability of the group had been substantially reduced as a result of an acquisition and subsequent restructuring of the clothing and fashion company.

It was established that despite an increase in turnover of approximately £100,000, the group suffered a net loss after exceptional items of expenditure during the year 2001. However, the company is a substantial and well-known fashion clothing company with stores located worldwide. The net assets of the group are limited for a group of its size and except for the exceptional costs arising on the reorganisation, the indicators are that the company has been trading profitably and turnover has been growing. There are no reasons to believe that it will not continue to do so.

Advice

I prepared a recommendation form for the proposed assignment to the client. This outlined the financial position of the proposed assignee, based on our assessment of the accounts, and gave advice upon whether to accept the assignment. I outlined the following points in addition to the accounts analysis:

- The assignee is a large company with 840 stores worldwide.
- They had six stores in the UK, four of which were outside London.
- Photographs of their existing stores were supplied, and I felt the new shop fit-out would improve the appearance of the building as a whole, and would add value. Attached to Appendix VI are photographs of an existing Soccerstores outlet (not reproduced here).
- Overall, the proposed assignee seemed to be a reliable tenant with good covenant strength.

My client accepted my recommendation to approve the assignment. Therefore, my client's main objective to secure an assignee with good covenant strength had been met, and thus the investment value had been maintained or enhanced.

Critical analysis

This section critically analyses the project.

My role in the assignment was to provide consideration to the terms of the lease, legislation, case law, the tenant's requirements and our client's best interests. I feel that I fulfilled this role by having regard to all of the above, and believe that my client was satisfied with the progress I made throughout.

The assignor first approached the client on 10 April 2003 seeking consent to assign. The assignor required a decision from the client, together with a licence to assign, by 2 June 2003. I had explained to the assignor that any pieces of information I requested would have to be delivered quickly as the process could be delayed if they were not. It was prudent to outline this, as it avoided any delays and ensured that the time scale given was met.

The assignor had only held the lease on the premises for a relatively short period of time (two years and eight months out of a 15-year lease.) Due to time restrictions and the wish by all parties to proceed with the assignment speedily, it was agreed between the parties that repairing and decoration covenants would not be enforced. This was my client's instruction, although when doing another assignment, I would look to ensure that the covenants are being complied with. In this case, the tenant covenanted to decorate the exterior of the premises every three years, and given that it was in need of decoration to the front façade, it would have been prudent to have enforced this covenant. I maintained a good relationship with the assignor and its agents throughout the transaction.

I have learnt a number of valuable lessons as a result of my involvement in this project. They can be summarised as follows:

The importance of careful analysis of the lease in determining the user and alienation provisions contained therein. I found this particularly important in the case of the alienation provision. It outlined circumstances where withholding consent to the assignment was acceptable. It also imposed conditions for the assignee and assignor to meet. It was important that both parties met these conditions as part of the assignment procedure, and important that I checked to ensure this was the case.

- I referred to the Code of Practice for Commercial Leases in England and Wales (2nd ed, April 2002) produced by the Commercial Leases Working Group 2002, to determine what it stated with regard to assignment in agreeing new leases. Recommendation 9 of the Lease Code, states that the restriction on assignment should be landlord's consent, not to be unreasonably withheld. I feel that some clients will be uncomfortable relying on the general concept of reasonableness, and will prefer instead to continue to list 'circumstances' or 'conditions' (many of which would be 'reasonable' anyway).
- My familiarity with the Landlord and Tenant (Covenants) Act 1995, Landlord and Tenant Act 1988 and Landlord and Tenant Act 1927, has increased, as I had to refer to sections from each of these Acts, as referenced in the lease document. My ability to apply statute to practice has therefore improved.
- My ability to work within tight time scales has focused me on identifying all pieces of information required to be asked for, and chasing up if such information is not provided in the depth requested. It was vital to all parties involved that the completion date of 2 June 2003 was achieved, and I managed to deliver within that time scale.

- Ensuring the file on the property is kept up to date and complete on such a contentious issue is important, especially if consent is refused.
- Ensuring the solicitors who are drafting the documentation are correctly briefed and understand the requirements of all parties (as they will have not been involved in considering the application and therefore may be unaware of important points).

Appendices

The Appendices were: Appendix I – Location Plan, Appendix II – Street Plan, Appendix III – Ordnance Survey Extract, Appendix IV – Photographs, Appendix V – Explanation of D&B Terms, and Appendix VI – Photograph of Existing Store. None of the Appendices are reproduced here.

Illustration of issues/possible interview questions

Q You said in your report that, 'The covenant strength of the proposed assignee would need to be the same or stronger for the client to accept the assignment'. Is this exactly what the lease said, or is this your interpretation of the clause?

Q What do you consider a reasonable/unreasonable time to be for responding to the tenant, and generally handling the assignment – and what factors may allow the landlord to have more time?

Q What are the consequences of delaying consent – ie assuming in another scenario that you acted for the tenant, how would you deal with a landlord seeking to delay matters, and what might be the basis of any claim for damages?

Q Are you aware of any case law on such matters?

Q How does the strength of covenant affect the capital/investment value, bearing in mind that the rent was the same?

Q Why did the tenant pursue assignment, and not sub-letting do you think?

Q Did you consider whether there may be benefit in the landlord accepting a surrender and granting a new lease to the assignee?

Q What is A1 – A1 of what and what other uses would fall in A1?

Q   Did you consider the scope for the landlord to secure a higher value use such as A3, and look to possibly negotiate a surrender with the tenant in order to secure a higher rent?

Q What planning issues prevented A3 from being a possibility?

Q Do you, in acting for the landlord, have an interest in any premium payable?

Q Incidentally, how are leasehold interests valued – how might you value the leasehold interest in this case if acting for the tenant who is seeking to assign the lease?

Q You mentioned side letters – what is the relevance?

Q Regarding the company accounts, what aspects did you analyse?

Q Why might gearing and cash flow be relevant, even if the business appears to enjoy strong profitability?

Q You referred to the commercial lease code in respect of assignment. Are you suggesting that a landlord should relax a requirement for an AGA even if the lease already contains a requirement for the tenant to enter into one?

Chapter 10

Investment Acquisition

This chapter examines an acquisition on behalf of a property investor of a London office investment. At the end of the critical analysis/report and appendices, an illustration is provided of other investment and financing issues.

The report is contributed by Rupert Mitchell.

Introduction

This report will focus on the valuation and acquisition of the long-leasehold interest of 56 Parkinson Street, London SW1. This instruction is typical of the type of work that I have been involved with during my training period, and illustrates many of the skills I have acquired.

For the duration of the project, I worked with a partner in the investment department. I can confirm that I have obtained authorisation from the client and my firm as to the contents of this report.

I will address the major issues encountered throughout the acquisition process, along with solutions and courses of action taken. I will examine the client's objectives, the way in which the asset was valued for purchase and the due diligence procedures. The report will conclude with a critical analysis of the outcome and a reflective summary of the experience that I have gained.

The instruction

A partner in the investment department was approached by a private investor to source and acquire suitable properties that met their investment requirements.

Following detailed discussions with the client, I felt confident I had a thorough understanding of their objectives. I knew:

- The reasons for purchasing an investment property.
- Preferred location(s).
- Preferred tenure and type of property.
- The preferred lot size and methods of financing.
- The required return.

I proceeded to review details of properties currently available to establish whether they broadly met the client's parameters. My search revealed an appropriate property.

Investment summary and initial findings

56 Parkinson Street was identified as a suitable opportunity. Located in the Victoria area of London, the building, dating from the mid 19th century, was

rebuilt behind a period façade in 1991. It comprises approximately 1,085 m² (11,687 sq ft) of offices, arranged on ground and three upper floors. The property is let to Alston Development Corporation at a passing rent of £520,000 pa, equating to £460 per m² (£42.75 per sq ft) in respect of the best quality office space. The demise also includes a three-bedroom town house known as 21 Marshall Street.

Detailed descriptions of the physical characteristics of the investment are attached at Appendix I (reproduced at p94). A Location Plan, Ordnance Survey Extract and Photographs are attached at Appendices II, III and IV respectively (not reproduced here). A summary of the Headlease and Underlease are included at Appendices V and VI respectively (reproduced at pp 96 and 98).

My initial enquiries suggested that this property was considerably overrented, as the Estimated Rental Value (ERV) was £406,950 pa, compared with the rent passing of £520,000 pa. A detailed summary of rental value is attached at Appendix VII (not reproduced here).

56 Parkinson Street broadly satisfied the client's objectives, providing the opportunity to acquire the long-leasehold interest of an office property located in an established location with an income stream secured against a single tenant for a further nine years on an institutional lease. The quoted purchase price equated to a Net Initial Yield of 8.50% – sufficient to cover loan repayments and management costs, while generating a return on the equity employed.

Having established that no conflicts of interest existed with this property, and ensuring that terms of engagement were in place, I drafted a letter of recommendation to the client summarising the property details, outlining the advantages/disadvantages of the investment and explaining initial thoughts on value.

It was important to remember at this stage that the rationale behind the acquisition was not that the purchase price represented Market Value (or better), rather instead that the main consideration was (given the client's objectives and methods of funding) that their requirements would be fulfilled through meeting expectations of worth.

Following an inspection of the property, I undertook further enquiries and discussed with the client my assessment of the worth of the property. Based on our initial recommendations and advice, the client instructed my supervisor to submit an offer of £5,415,000, subject to contract, surveys and the usual due diligence. This price reflected a Net Initial Yield of 9.07%, and an equivalent yield of 8%. This offer was below the asking price, but I was of the opinion that this was a true reflection of the property's value.

The offer was accepted, and I was responsible for the reviewing the Heads of Terms. Throughout this period, I was responsible for liasing with the other parties involved – predominantly the vendor's agents and the purchaser's solicitors.

Having agreed Heads of Terms, I went onto undertake more detailed enquiries and prepare a purchase report. Initially I requested, in writing, confirmation from the client of the basis of valuation of the property for the purchase report. In doing so, I was complying with the RICS Appraisal and Valuation Standards 5th ed (the 'Red Book').

Key issues and options including proposed solutions

To ensure the client was achieving best value, I conducted further research on matters that could impact on value, either now or in the future. The major issues that arose are discussed below, along with reasons for the rejection of certain options and the solutions I arrived at.

Comparable evidence

Due to the weak state of the occupational market at the time of acquisition, it was vital to ensure accurate up to date evidence was collected and analysed.

My searches, including discussions with local agents, uncovered few comparables in the immediate vicinity. However, one particular transaction was revealed. The first floor of 28 Mallinson Terrace, comprising 104 m² (1,118 sq ft) had recently been let at £376.70 per m² (£35 per sq ft). This was located reasonably close to the subject property, but of slightly inferior specification – and, additionally, the floor areas were not comparable. As my initial investigations had realised few comparables, there were a number of options I could pursue:

- Search for more historic transactions. This option was ruled out as the state of the occupational market at this time was such that rental values were falling, therefore comparable evidence was becoming quickly outdated, and not a true reflection of the market.
- Search outside the immediate vicinity. I pursued this option and various relevant comparable transactions were revealed, including a recent letting at 4–6 Mercantile Gardens, at £35 per sq ft. However, the rent review provisions of the subject property specified an assumption of a 10-year term, and given that this evidence was for a five-year term, an allowance had to be made.

Other evidence at some of the best buildings in Victoria indicated a prime headline rental level of £47.50 per sq ft.

Although the lease did not specify that the rent review mechanism permitted deductions for any incentive periods offered in the market, void and rent-free periods were also studied in order to give an accurate picture of the market. A summary of rental evidence collected is attached at Appendix VIII (not reproduced here).

Having considered the evidence and discussed the property with my office agency colleagues, I confirmed the Estimated Rental Value of the property to be in the order of £32.50 per sq ft. The rental value had therefore fallen considerably from the £42.75 per sq ft set at the last rent review.

In addition to research into the ERV, rental growth/decline predictions over the coming years were also studied, along with the subsequent effect these would have on the property. These findings were reported to the client.

Given the abundance of recent transactions, it proved relatively easy to arrive at an opinion of Rental Value for the residential element. I did not consider capital values due to the restrictions imposed by the Headlease.

Next I searched for relevant comparable transactions of investment sales. Along with the regular considerations, particular attention was paid to a number of other factors, including the unexpired term of the long leasehold interest. In addition to this, the existence of a restriction to residential use within part of the demise meant some investors may view the property to be an inferior product. The tenant did have a relatively weak covenant, however, I proved there to be good demand in the investment market for good quality office buildings in core locations.

A number of transactions were researched for comparative purposes. Of particular interest was the recent purchase of the long-leasehold interest of 29–31 Mercantile Gardens. Let to Switchtime for a term of 15 years, the property was rack-rented with a purchase price representing a Net Initial Yield of 6.75%. By comparison, the subject property was overrented with a shorter (occupational) lease length and a weaker tenant covenant strength. The leasehold interest incurred a rent payable to the freeholder of 15% of market rental value, resulting in a less attractive investment profile.

Another recent investment transaction was that of 10 Canterbury Street. The property, held freehold, was let to International Bank with 11 years unexpired. It was over-rented by around 23%, and was purchased off a Net Initial Yield of 6.57%, and an equivalent yield of 5.80%. Although benefiting from a stronger covenant and held freehold, the property offered a basis to which the subject property could be compared. Detail of these and other transactions are summarised at Appendix VIII (not reproduced here).

In arriving at an appropriate yield to apply to the income, other transactions were analysed and discussed with colleagues who benefited from more experience in the market.

Covenant strength

I commissioned a Dun & Bradstreet report to facilitate a greater understanding of the tenant's covenant strength. The findings of the report indicated a rating of 'O2', reflecting a financial strength which is undisclosed, but a risk factor representing low risk. Further research, including obtaining a set of recent accounts, revealed that the covenant of the tenant was not as strong as previously believed. Given all the income was derived from this tenant, combined with the element of overrenting produced a less secure 'top-slice' of income, resulting in a less attractive investment profile.

I provided our view of the market perception of the tenant covenant strength, but I advised the client that I was not qualified to give financial advice on covenant strength, and they must ultimately satisfy themselves in this regard (which they did, arriving at similar conclusions). Given that the client intended to purchase the property partly through debt, this was an important consideration.

Having discussed these findings with the client, it was concluded that given the nature of the portfolio, an investment of this nature carrying a heightened level of risk was acceptable in order to increase the 'running yield' of the portfolio as a whole.

Method of appraisal

Recent comparable transactions provided me with a basis on which to value the property. Although the letting market was fairly weak at the time of purchase, the investment market was proving to be strong for good quality buildings in core locations, with prime West End yields in the order of 5.50%–5.75 %

The fact that the property was held on a long-leasehold basis was a key consideration in the valuation process. One factor that often detracts from the attractiveness of an investment is an onerous level of gearing. In this instance, this was not a problem as the gearing was a nominal sum. However, with an unexpired term of 111 years, the property was a wasting asset, because as the unexpired term diminishes and the asset's appeal to an investor reduces, the value will fall. Through my research and other market knowledge, I concluded that the investment market is sensitive to the psychological barrier of less than 100 years unexpired.

As a result of my findings, in order for the client to maintain the value of the investment, I recommended at some point they may consider opening a dialogue with the freeholder to explore the possibility of extending the lease or acquiring the freehold interest.

In order to establish our opinion of market value, I adopted the investment method of valuation having regard to the available evidence and wider market knowledge. After considering the various methods of valuation such as the 'hardcore and topslice' ('froth') method, the equivalent yield method was chosen due, in part, to the amount of relevant, comparable transactions available at the time of reporting (discussed earlier and detailed at Appendix VIII – not reproduced here).

Given that there is rarely a perfect comparable transaction, the adopted yield reflected a combination of a yield discount to properties with more secure income and/or better positioned to benefit from rental growth. In particular, a yield movement of + 0.75 % – 1.00% was noted on those properties held long leasehold as opposed to a freehold basis. Due to the fact the property was held long-leasehold, and in keeping with market practice, I adopted a sinking fund (at 4% and 30% tax) in my valuation.

My valuation, at a Nominal Equivalent Yield of 8.00%, producing a Net Initial Yield of 9.07% (after allowing for purchasers' costs of 5.75%) proved to be fully supportable in the market place. A copy of the valuation is attached at Appendix IX (not reproduced here).

Due to the level of overrenting it was considered that the prospects for any rental growth above the passing rent on this property were slim. Therefore particular regard was had for the level of Net Initial Yield as this would be the return at least until the next rent-review, and most likely for the remainder of the term.

Inspection, measurement and surveys

With my supervisor I undertook a measured survey on 10 April 2003. Both the Headlease and the Underlease specified an area of the property restricted to residential use. In accordance with the RICS Code of Measuring Practice (5th ed) we therefore measured the office element of the property on a net internal basis, and the residential element on a gross internal basis.

I recommended to the client that a building, mechanical and electrical survey should be commissioned to establish the condition of the building and its services.

The most significant finding of the mechanical and electrical survey was a few minor defects in the air-conditioning system. I requested the client's solicitors confirm that the drafting of the tenant's repairing obligations in the lease were sufficient to cover these items of disrepair and maintenance, who duly confirmed that it was. This course of action was chosen to ensure that our client's income would not be eroded by any expenditure on these items of disrepair, and to ensure that there would be no detrimental impact on value. These findings were discussed with the client.

Statutory enquiries

As part of the due diligence procedure, I carried out research into various planning issues. From my verbal enquiries of Westminster City Council I noted that the planning officer believed an 'in principle' change of use to residential would most likely be approved. I therefore undertook an appraisal to establish residual residential value should vacant possession be obtained. The residual value arrived at was similar to the agent's quoting price. However, the constraints of the Headlease meant that conversion to residential was effectively ruled out without the participation of the freeholder. This was reported to the client who was satisfied that the property benefited from underlying residential value.

Landlord and Tenant

I obtained the relevant legal documentation in order to verify the information provided by the vendor's agent. This included preparing lease summaries of both the Headlease and the Underlease, in particular noting:

- The existence of any rights reserved or granted over the property.
- Alienation provisions, paying particular attention to the restrictions on the assignment of the residential element of the property.
- The rent review provisions.
- Repairing and use covenants.
- Any other factors that were likely to impact on investment value.

It was noted from the Headlease that the alienation provisions prohibited an assignment of part only of the demise. The consequence of this restriction was that the residential element of the property could not be sold off separately at some later date if so desired. This restriction was discussed with the client and factored into the valuation. This course of action was taken so as to ensure that the client was advised of the restrictions they faced upon purchasing the property.

The relevant/value sensitive factors are annotated in the lease summaries at Appendix V & VI.

Critical appraisal of the outcome

Throughout this project, I encountered a variety of professional, technical and practical issues.

The comparable evidence collected was sufficient to feel confident that I had not omitted a recent deal of significance. However, with more time it would have been useful to undertake further research in the Victoria office market.

It would also have been beneficial to inspect the property with a colleague from the Office agency department. Although I discussed the property with them at length, providing a full description including photographs, it would have been useful for them to actually see the property when arriving at their opinion of ERV.

As discussed previously, I made comprehensive oral enquiries of the council's planning department which I recorded by way of file notes. In retrospect I should have written to the planning department so that I had correspondence on file. This did not, however, materially affect the outcome.

As with many acquisitions, time was in short supply, therefore my time management skills were tested and I had to organise myself efficiently. I had to work to a tight time scale to ensure that the client's objectives were met while still following the procedures involved within the acquisition and due diligence processes.

Summary of experience gained

This project has helped the development of my common/mandatory, core and optional competencies. In addition, I believe that the market knowledge gained throughout the project has benefited me and will be of use in the future.

Common, mandatory competencies

This project involved regular contact with the client, vendor's agent, client's solicitor and other surveying colleagues. This interaction has developed my ability to play an active role in discussions, seeking and providing advice where appropriate.

Core competencies

This project has continued the development of my valuation experience. Given the lack of directly comparable evidence in the immediate vicinity, I had to search for evidence in the surrounding areas, and consequently gained experience in making the necessary adjustments to comparable evidence. In addition, I inspected the subject property, taking full site inspection notes, supported by photographs.

Optional competencies

I have furthered my understanding of landlord and tenant issues, in particular when dealing with long-leasehold interests. This project highlighted the importance of undertaking accurate and thorough lease summaries, as well as the value of discussing issues in the lease with the solicitor.

In terms of Property Marketing, this project gave me further insight into the acquisition of commercial property. In particular, I have learnt the necessity of thorough pre-acquisition due diligence, which provides the opportunity to analyse any factors influencing valuation prior to the acquisition.

With regard to Development Appraisal, although this property was purchased as a standing investment, the fact that there was a significant alternative use value resulted in my skills in this competency also being practised.

I have also gained vital experience in another of my optional competencies, Property Investment Funding. The importance of debt for the private investor was highlighted, as well as the relationship (and profitability) of such funding to the strength of covenant.

Other professional skills that I feel I have benefited from during this project include:

- Understanding and following client's requirements.
- Being aware of issues on site, including natural light, layout and reception areas.
- Taking site inspection notes and calculating floor areas.
- The valuation of long leasehold interests.
- Report writing.

I have learnt a considerable amount about the level of service required by the client and the need to provide clear, concise advice which can only be achieved through adequate research, analysis and preparation. This was illustrated when collating market evidence where accuracy is paramount.

To conclude, this project has given me the opportunity to work on my own initiative, while still being supervised by a qualified surveyor. This has increased my confidence to carry out future instructions both competently and professionally.

Appendices

The Appendices were Appendix I – Summary Details, Appendix II – Location Plan, Appendix III – Ordnance Survey Extract, Appendix IV – Photographs, Appendix V – Lease Summary – Headlease, Appendix VI – Lease Summary (underlease), Appendix VII – Summary of Rental Value, Appendix VIII – Comparable Schedules and Appendix IX – Valuation Print-Out.

Appendices I, V and VI set out below illustrate the descriptive information included in a valuation/purchase report, and also lease terms and their possible implications.

Appendix I: Summary details – 56 Parkinson Street, London, SW1

Location

The property is situated in Victoria, which is within the City of Westminster, in the heart of central London. Victoria is located immediately to the south of Mayfair and St James's (which are the prime West End office locations), and approximately two miles to the west of the City of London.

The Victoria area benefits from excellent public transport links. Victoria mainline and underground stations are located close to the subject property. The mainline station provides local and regional rail connections to the south of England, while the nearest London Underground station is Victoria (Central, District and Victoria Lines) linking the area to the rest of London. In addition, St James's underground station (Circle and District Lines) is also situated close by to the east of the property. Victoria mainline station also provides direct links to Gatwick Airport, with the journey time on the Gatwick Express of approximately 30 minutes.

Situation

The property is situated at the junction of Parkinson Street and Marshall Street, mid-way between Mallinson Road to the north and Edward Street to the south.

The property occupies a prominent position, close to the commercial hub of Victoria. The Victoria office market is attractive to major office occupiers because of the number of large, modern office buildings, its proximity to Westminster and government departments and its central location within London. It benefits from good public transport links to all parts of central London and to the London Airports.

Directly opposite the subject property is the site of a major new office and retail development.

The area also contains a number of residential properties and comprehensive shopping facilities on Edward Street and around Victoria Station.

Description

The subject property was originally constructed as a purpose-built church, known as the Church of Paul and Matthew, during the middle of the 19th century. The original development incorporated a presbytery building on the corner of Parkinson Street and Marshall Street constructed integrally within the church.

A comprehensive redevelopment of the property was undertaken in 1991 to provide approximately 11,700 sq ft (1,086 m²) of office accommodation within the converted church structure, while the presbytery has been converted to a self-contained three-bedroom residence known as 21 Marshall Street.

The property is Grade II listed and consequently the original solid brick elevations to the buildings have been retained. As part of the redevelopment, an internal steel-framed structure has been constructed to enable a modern, relatively column-free, open plan floor plate.

The property is presented in good decorative order throughout. In terms of specification, the office space is served by a four-pipe fan coil air conditioning system, suspended ceilings and raised floors.

The building also benefits from male and female wcs on alternate floors, a staircase at both the eastern and western end of the floor plates, and an eight-person passenger lift.

There are seven car parking spaces located in the basement that is accessed via a car lift from Marshall Street.

The residential town house accommodation benefits from its own self-contained access from Marshall Street.

Site

The property occupies a site that comprises approximately 0.048 ha (0.118 acres).

Floor areas

The property has been measured in accordance with the RICS Code of Measuring Practice (5th ed) and provides the following approximate net internal floor areas:

Floor	Use	Area (m^2)	Area (sq ft)
3rd	Offices	248.33	2,673
2nd	Offices	289.20	3,113
1st	Offices	285.12	3,069
Gr	Offices	245.35	2,641
Gr	Reception	17.74	191
Total		1,085.75	11,687

Appendix V: Lease Summary – Headlease

Lease date

7 February 1989.

Lessor

The Westminster Trustees.

Lessee

Construction Properties Ltd.

Premises

All that land and building formally known as the Church of Paul and Matthew now known as 56 Parkinson Street and 21 Marshall Street, London SW1.

Term

125 years from and including 7 February 1989.

Rent

£675.00 pa. This rent comprises £575 pa for the office accommodation and £100 per annum for the residential accommodation.

Rent review

The rent for the residential accommodation is reviewed every 30 years, and will increase by £50 at each review. These uplifts are specified within the lease under as follows:

0–30	£100
30–60	£150
60–90	£200
90–120	£250
120–125	£300

The rent for the office accommodation is fixed for the remainder of the term.

User

The basement, ground and upper floors of the former presbytery shall be used as residential accommodation, and all that part of the premises being the former church of Church of Paul and Matthew of the former resbytery shall be used as office accommodation.

This restriction on use, coupled with the alteration provisions, essentially means that the property could not be converted to provide all office/residential accommodation without the participation of the landlord.

Alienation

The tenant is not permitted to assign part only of the premises. However, the tenant may assign the whole or under let the whole or part of the premises with the prior consent of the landlord, such consent not to be unreasonably withheld. Prior to any assignment, the assignee must enter into direct covenants with the landlord to perform and observe all of the tenant's covenants. Each and every permitted underlease of the office accommodation cannot be granted at a rent less than the open market rental value of the office accommodation.

These provisions effectively restrict the long leaseholder from selling off the residential element of the property.

Insurance

The tenant covenants to insure the premises for full reinstatement value, including professional fees and other incidental expenses, including third party and occupiers liability insurance.

Alterations

The tenant is not permitted to make any external alterations or alterations to the premises without the prior written consent of the landlords (such consent not to be unreasonably withheld).

Appendix VI: Lease summary – underlease

Lease date

8 May 1997.

Lessor

Principal Assurance plc.

Lessee

Alston Development Corporation.

Premises

The building known as 56 Parkinson Street Street and 21 Marshall Street, London SW1.

Term

15 years from 25 March 1997, expiring 24 March 2012.

Rent deposit

£162,500.

Rent passing

£520,000 pa, as from March 2002.

Rent review

Upwards only every fifth year, the next review being 25 March 2007. The rent is to be reviewed to the yearly rent for which the demised premises might reasonably be expected to be let on the open market, with vacant possession for a term of 10 years. The usual assumptions and disregards apply.

User

To use the town house as residential and the remainder as offices with car parking at basement level. However, there is an express provision not to use the premises as the following:

- Betting office.
- Residential estate, travel or employment agency.
- Unemployment agency or benefit office.

- Offices to which members of the general public have unsolicited access to the premises.

Repair

The tenant covenants to keep the demised premises in good and substantial repair and condition, and to yield up at the termination date in accordance with the covenants by the tenant as contained within the lease. There is also a covenant for the tenant to decorate internally and externally in the year 2000 and in every fifth year, the term thereafter and in the last six months of the lease.

Alterations

The tenant is prohibited from construction and new building or new structures of any kind, and to make any structural, internal or external alterations or additions to the premises, other than with the prior written consent of the landlord (not to be unreasonably withheld). The tenant can carry out internal non-structural alterations to the premises without landlord's consent.

Alienation

The tenant may assign, underlet or charge the whole of the premises with the landlord's prior written consent (not to be unreasonably withheld).

The tenant is permitted to underlet the whole of the residential part and a whole floor or two or more contiguous floors of the premises (excluding the residential part) together with pro rata proportion of the car parking spaces with landlord's consent.

In the event of an assignment, the current tenant shall enter into an Authorised Guarantee Agreement. If the landlord reasonably requires the tenant will retain one or more acceptable guarantors for the proposed assignee, a reasonable rent deposit arrangement and/or provide such additional security for a performance by the proposed assignee of its obligations.

Insurance

The landlord covenants to insure the building for reinstatement value and the loss of rent for three years or for a longer period that the landlord may from time to time reasonably consider to be sufficient for the purposes of planning and carrying out any such reinstatement. This premium is recoverable from the tenant.

Illustration of issues/possible interview questions

Q How much did the client have to spend, and did you consider whether it would be preferable to acquire a number of properties, especially given the risk of overrenting and covenant strength?

Q 56 Parkinson seemed to be quickly established, but what other investment properties did you find, and what were their investment attractions?

Q Can you please remind us what was the impact of the residential element on your valuation, and did the residential element give rise to any particular rights?

Q Could the residential element be sublet, and any rights created for a tenant which may be detrimental to the acquiring client's interest at any stage in the future?

Q What in your experience is the view of lenders to long leaseholds of such properties?

Q How would a lender structure a loan bearing in mind that there is nine years to lease expiry?

Q How might the fact that the property is more than 25% overrented influence the terms of finance achievable?

Q How was the Net Initial Yield calculated?

Q Can you explain why a high yielding investment (albeit high yielding due to overrenting) can be attractive for an investor looking to build a property investment portfolio?

Q You mentioned your offer reflected a Net Initial Yield of 9.07%, and an equivalent yield of 8%. How did you calculate the equivalent yield?

Q You referred to hardcore and topslice. Was this how you valued the property, and if so, what might an alternative be?

Q Talk us through the key elements considered in the company accounts.

Q Regarding the leasehold nature, why was a sinking fund adopted?

Q Is this market for practice for leases of over 100 years to expiry, and what, approximately, would be difference have been between an 8% yield applied to a freehold, and a 8%, 4%, 30% yield applied to a leasehold?

Q Did you seek to establish whether the vendor's agent had any other interest?

Q Were you surprised that the first offer was accepted/was there any way that the price could further be reduced?

Q Can you please clarify the basis of the rent review in the underlease – how will the rent take account of a rent-free period as an incentive, and a rent-free period as a fitting out allowance?

Q Can you recall four cases from the 1990s which could be researched to give further clarification on this point regarding headline rents at rent review?

Finance/property investment illustration

A brief illustration is shown below of how the ability to raise finance influences how property investors can make money, the following is provided by Midlands Property Training Centre, based on issues covered in training sessions.

This shows that the yields considered as part of investment valuation work are only a component part of the property investment process.

If an investor has £100,000 to invest, they may acquire a property for £100,000 cash, and find a tenant who pays a rent of £10,000 per year. The return is 10% pa in a similar way that a bank or building society account may earn, say, 4% interest*.

If the investor decides to use the £100,000 cash as their deposit/equity for a larger property instead, and borrow money, the bank may provide a £300,000 loan. This would be a loan to value ratio of 75% (£300,000 divided by £400,000 – ie commonly termed a 75% mortgage in the residential mortgage market). If a 10% return was available, and the cost of finance/the interest repayment rate was 7%, the profitability of the investment would be as follows (ignoring costs):

Cash element	£100,000	10% rent		£10,000
Borrowed element	£300,000	10% rent less 7% finance	£30,000 £21,000	£9,000
				£19,000

£19,000 ÷ £100,000 = 19% return.

The ability to borrow almost doubles the investor's income return from 10% to 19%.

(The 10% is really the all risks yield that would have been used in an investment/capital valuation.)

Also, if for example, property values went up by 25%, investing only £100,000 cash and not borrowing/not buying a larger property would produce a profit of £25,000/25% on the initial outlay. But if using the £100,000 to buy £400,000 worth of property, a 25% increase in property values would be £100,000 – ie doubling the initial outlay.

The increase in the capital value of the property to £500,000 from £400,000 could allow the investor to sell and take profits, but could also facilitate refinancing. In preserving the LTV of 75%, the bank would provide the investor with £75,000 (75% of £100,000). At an LTV of 75%, this could be used to acquire a property of £300,000. The process can go on and on, and see the investor quickly build a portfolio, with other sources of finance also being drawn on.

However, risk also needs to be considered. Poor market conditions could see rents fall, tenants depart and property prices/capital values fall. The same sensitive effects which increased returns could increase losses.

---

* Note however the differences between a property investment and a bank account investment – the property investor may benefit from an increase in the value of the property/their initial capital, whereas the bank account deposit will remain fixed. Rents and interest rates may indeed rise and fall, but rents will generally rise upwards over time, whereas interest rates will vary within a range.

Chapter 11

Insolvency and Sale of Property

This chapter provides an insight into the specialist field of insolvency where in disposing of a property, the surveyors' role has particular regard to the relevant legislation. Development appraisal is also included. The report is contributed by Kieran McLaughlin.

Introduction

This report outlines my involvement in the management and disposal of the Law of Property Act (LPA) 1925 Receivership of a former residential care home facility Preston.

I have been involved with a number of LPA Receiverships over industrial, office and retail properties and also specialist property types which are often more valuable as an alternative use. Therefore, I feel that this project is a true and accurate representation of the type of work in which I have been involved in throughout my training period

Summary of process

Outlined below are the issues to address through the life cycle of the LPA Receivership.

- Ensuring valid acceptance of LPA appointment.
- Property inspection and collation of information.
- Initial LPA Receiver's Report (informal opinion of value, strategy to realise security).
- Informal valuation.
- Marketing, including selection of disposal method.
- Resignation.

These form the main structure of my case study.

Initial instruction and client's objectives

On 15 December 2002, a bank gave notice to my firm of the possible appointment of LPA Receivers over a former residential care home facility in Preston as a result of the mortgagor defaulting on loan repayments.

The client's objectives are to realise the asset as a result of a defaulting loan. The mortgagee appoints LPA Receivers to avoid the onerous liabilities attaching to a mortgagee in possession.

The essential difference between mortgagees in possession and LPA Receivers is that while the mortgagor acts on its own behalf, the LPA Receiver, although appointed by the mortgagee, acts as agent for the mortgagor. Therefore, appointing an LPA Receiver has the advantage for a mortgagee of removing the property from the mortgagor's control without adopting the associated liabilities.

The property

The property comprises a two-storey residential care home converted from a former school house which provided accommodation for approximately 20 residents with self-contained manager's quarters.

The property is of brick construction with pebble dash render under a pitched slate tiled roof. Internally, the property comprises an entrance hall, common rooms, kitchen, disabled toilet and 16 bedrooms on the ground and first floor. Externally, there is a garden to the north of the property, and the boundary of the site is identified by hedgerow and dwarf walling. Car parking is limited to on street parking.

Photographs of the property are included in Appendix I (not reproduced here).

The approximate gross internal floor area is 589.45 m^2 (6,345 sq ft. and the building sits on a site of approximately 0.09ha (0.22acres).

A plan of the site is included in Appendix II (not reproduced here).

Conflict of interest

I carried out an initial check to ensure that there were no conflicts of interest involving the subject party or any of the parties concerned in order to satisfy my firm's in-house Quality Assurance procedures (ISO 9002). This check was completed by performing a number of searches on my firm's instruction database. No potential conflicts were detected and the bank was notified accordingly.

Ensuring valid acceptance of the receiver's appointment

On 20 December 2002, (names of colleagues) were appointed as Joint LPA Receivers under the terms of a mortgage given by the mortgagor to the mortgagee.

It is imperative that solicitors ensure that the appointment is valid as soon as possible because if the appointment is deemed invalid, then the Receivers could be liable in tort as trespassers to the mortgagor's property.

Our appointed solicitors confirmed that the joint receivers were validly appointed as LPA Receivers, or receivers appointed under the terms of fixed charge. The solicitor's report also clarified the extent of the bank's charge, and confirmed that there were no problems on title, and no existing second charge holders.

The powers granted to the receivers can vary depending on the mortgage document; however the solicitors confirmed that the receivers had the benefit of all the statutory powers conferred under the Law of Property Act 1925, together with the powers expressly set out in the charge. This included:

* The power to borrow.
* The power to carry out repairs.

- The power to collect rents.
- The power of sale.

Although the LPA Receiver can now continue to exercise his powers once valid appointment is confirmed, it is important to realise how the receiver's position could change following the appointment of a liquidator over the company or bankruptcy if it concerns an individual. If the mortgagor goes into liquidation/bankruptcy, the LPA Receiver can no longer act as its agent, as the agency relationship will be terminated and will then need to be appointed as agent to the mortgagee to retain any element of control.

Notification of appointment

On the day of appointment, I arranged for the necessary letters of notification to be sent out to relevant parties advising of the appointment.

The letters and their function are summarised in Appendix III (reproduced at p113).

NB. The LPA Receivers are obliged to ensure that the property is properly insured. On the morning of appointment, I contacted the insurers to place the property on cover, and followed this up with written correspondence. A copy of the completed questionnaire is included in Appendix IV, and a copy of the Reinstatement Cost Assessment is included in Appendix V (not reproduced here).

The recognised procedure is that the insurers forward a property questionnaire requesting information and a reinstatement valuation. In order to protect the receiver's position, a second opinion is usually obtained from a qualified building surveyor for the provision of the reinstatement figure.

Initial inspection and collation of information

On the date of appointment, I attended the property with a qualified surveyor and under supervision, conducted a full measured survey and inspection of the property in accordance with the RICS Code of Measuring Practice. Particular regard was made to the following:

- The location.
- A brief description of the property including noting the age and construction.
- Physical condition.
- Potential liabilities under environmental legislation.
- Services.
- Comparable evidence search.

On my return to the office, with the aid of my notes taken at the time of inspection, I calculated the floor areas on a gross internal basis. Then, I carried out a number of enquiries, and was able to determine the following:

- The property had a domestic tax rating of Band 'E' and was not listed on the local authority's register of contaminated uses. Given the historical use of the property as a former school-house and residential care home, I was of the opinion that the property did not present any significant environmental risk.

- The property was accessed by an adopted public highway, and there were no other planned highway proposals that could have had a potential impact on the property. The property had consent for its use as a residential care home facility and the immediate area was unallocated under the local adopted Development Plan. Although the property was constructed in 1900, I was advised that the property did not have any listed building status, nor was it included in a Conservation Area.

- After meeting with the local planning authority, I found that the planners would, in principle, favourably consider other forms of residential use and did not attach any special interest in preserving the existing building. Alternatively, other uses such as a community recreational facility or a crèche/day centre would also be acceptable. (I already had an enquiry from a children's dance school wishing to relocate, and the planners considered this an appropriate use.)

Initial LPA Receiver's report

Each lending institution has different requirements and reporting time scales following the appointment of LPA Receivers. Debt recovery managers are not interested in a lengthy detailed description, but are more concerned with the problems, solutions and options, ie:

- Sell as soon as possible (private treaty or auction?).
- Seek planning for alternative use and therefore delay sale.
- Rent out and sell as investment.
- Undertake works to increase net realisation.

Similarly, debt recovery managers do not require the expense of a formal Red Book valuation, and an informal opinion of value will usually suffice. However, I still had regard to the Red Book when undertaking the valuation as a guide to best practice. The initial LPA Receiver's report highlights potential problems and pitfalls, and provides a marketing and realisation strategy encompassing an informal opinion of value.

Informal opinion of value

In order to ascertain the Market Value of the property, I sourced comparables from Estates Gazette Interactive, the Focus Database, the Ei Group (auction results) and contacted local property agents active in the Preston market.

The most relevant comparable was Ashton House, Preston, a 32-bedroom former care home set on 0.94acres which was over three times larger than the subject property. This sold for £123,000 at a local auction on 4 December 2002 after previously selling for £75,000 on 19 July 2002 at a London auction house. This comparable and the subject property were both located in secondary areas, and had been subjected to vandalism. Initial indications therefore suggested that the value of the subject property may be under £50,000 if analysed on a pro-rata basis against this comparable.

More historical comparables include the sale of a former care home in Liverpool in late 1999 for £110,000, and also another care home in Llandudno at the same time also for £110,000. Both these properties were offered for sale by private treaty with vacant possession, and were advertised as being suitable for conversion to other residential uses. (The details of these comparables are included in Appendix VI.)

The subject property was run-down internally, having been vacant for three to six months prior to the appointment of LPA Receivers, and required a complete refurbishment to return the property to its use as a residential care home. I enlisted the assistance of my firm's healthcare department to discuss marketing strategy and value. Upon investigation, they thought it highly unlikely that the property would appeal to the healthcare market given its location and condition.

The property was converted to a care home facility in the 1980s, and the required standards of accommodation have improved since then. The minimum room size now permitted is 9.3 m² (100 sq ft) and a number of the rooms were below this threshold. The necessary physical modernisation required to obtain re-registration would have reduced the operating capacity by 25%.

Having regard to the Red Book, the correct valuation basis would be 'market value of the empty property having regard to trading potential'. I was unable to obtain any previous trading accounts to assess profitability, and as the bank's charge did not extend to the chattels, the property could not be valued as a fully equipped operational entity.

The valuation having regard to trading potential would have required assumptions in relation to costs of refurbishment to satisfy room size requirements, provision of all chattels, costs of re-registration, income, and wage levels. The extent of the assumptions required would have made this valuation approach highly subjective. On this basis, the best indication of value of the property as it stood was the comparative approach in relation to the aforementioned comparable.

The property has restricted site circulation with street parking only, and is surrounded by high-density low value terraced housing. When trading, the facility could accommodate circa 15 residents, but the larger healthcare operators would not be interested in a property of this size, as they would be unable to avail themselves of the necessary economies of scale in service provision. The quality of location would also not fit their business model. The subject property had simply become physically and economically obsolete in relation to its former function. Therefore, I was of the opinion that the property had a nominal value if valued as a residential care home.

As previously stated, the planners had given positive feedback to the idea of some form of residential use given the character of the surrounding area. I then discussed the property with our in-house building surveyors and residential development department to discuss build costs/refurbishment costs and sales prices. After undertaking residual appraisals considering a number of different scenarios, I was of the opinion that the market value of the property was in the region of £50,000 to £75,000. A copy of the residual valuations considering site clearance and conversion is included in Appendix VII (reproduced on p112).

My calculations also led me to believe that the property was worth more as it stood as opposed to site clearance, which became an important factor in my disposal strategy.

Strategy to realise the bank's security

I discussed my findings with the LPA Receivers and expressed my concerns and recommendations. I thought that the property was worth more on an alternative use basis through refurbishment of the existing structure, rather than site clearance (which is supported in Appendix VII). I considered the possibility of involving our planning department and gaining outline planning consent for a residential scheme before taking the property to the market, but my strategy was heavily influenced by other factors.

In the days leading up to our appointment, the property was broken into on two occasions, and given the nature of the building and the surrounding area, I was of the opinion that it was extremely prone to break-in/vandalism/arson.

Having weighed up the holding costs and the high security risk, I considered that the net realisation would not have significantly improved if outline planning consent was sought. The decision was taken to place the property on the market immediately.

In relation to the receiver's duty of care, case law has determined that the receiver is obliged to carry out positive steps to maintain value, but there is no requirement to take any steps or incur expenditure to enhance the net realisation. It has been successfully argued that the market will reflect the likelihood of obtaining planning permission through hope value when bidding for the property. A brief synopsis of the Silven Properties case regarding the receiver's duty of care has been included in Appendix IX (not reproduced here).

I then considered whether to sell by private treaty or through auction. I weighed up the pros and cons of each method of sale, and thought that auction was preferable. Through auction, the LPA Receiver can demonstrate that the property has been fully exposed to the market by an independent party, thus limiting personal liability while achieving a timely realisation of the asset. Holding costs for the property including insurance, interest accruing on the loan and security patrols were estimated to cost £1,500 per month, and therefore it was advantageous to give the bank a relatively certain exit date – as opposed to the indefinite marketing period associated with sale by private treaty.

The residential market is currently very buoyant, fuelled by the prevailing low interest rates. The surrounding area is made up of relatively low value terraced housing and consequently this would hamper the gross development value of any proposed residential scheme. Therefore, I was of the opinion that the property would appeal mostly to local developers who concentrate on smaller lot sizes. This type of buyer generally acquires properties off-market or at local auctions and I was confident that a North West regional auction, rather that a national auction, was more appropriate to flush out all interested parties. I feel that this was demonstrated with the sale of the aforementioned comparable which sold for £48,000 more at a local auction than it did six months earlier at a London auction. I believe this property simply missed its target audience at the London auction house.

I then arranged for the property to be entered into an auction which was due to take place on 26 February 2003. The auctioneer was appointed by the LPA Receiver on a joint agency basis. I advertised the property in the local press and in a Healthcare Trade journal where businesses are frequently offered for sale, and also drafted a marketing brochure in accordance with the Property Misdescriptions Act 1991. I decided to do this after discussions with the LPA Receiver to ensure maximum exposure to the market through a relatively cost-effective medium. A copy of the marketing brochure and newspaper advertisement are included in Appendix VIII (not reproduced here).

Prior to the auction, I received a number of bids ranging from £50,000 to £100,000, which I relayed to the receiver and the appointing bank. However, it was decided that the property should go to auction as planned.

As a result of the high level of interest and bids received, I decided to set a guide price of £60,000 to £70,000 and an undisclosed reserve of £85,000. This guide was set to generate interest while the reserve allowed the LPA Receiver to keep control of the sale – ie if the reserve was not met, I would have been free to approach the party that initially offered £100,000 after the auction.

I attended the auction, and in keeping with the terms of the auction, contracts were exchanged on the day for £140,000 with completion to follow within 28 days thereafter.

As the property was sold without planning permission, it was open to wide interpretation as to the proposed use, the density of development and the standard of accommodation to be offered. These variables make accurate valuation extremely difficult, and also do not include the possibility of a special purchaser in the market.

Managing the receivership

In accordance with the Insurers' Code of Practice manual for vacant properties, I arranged for the windows to be boarded up at ground-floor level and the gas, water and electricity to be disconnected where necessary. A local security firm was also contracted to undertake drive-by inspections four times a week.

Resignation of LPA Receivers

Following completion of the sale of the property, the LPA Receivers were in funds to repay the overdraft facility which was used for marketing, insurance, security, auctioneer's fee and solicitors' fees. The LPA Receivers' fees were deducted from the sale proceeds and net funds distributed to the bank, together with a full income and expenditure report.

I arranged for letters to be sent out to the necessary parties, advising that the LPA Receivers had disposed of the property on 19 April and had now ceased to act.

Reflective analysis

My key objective was to implement a strategy and manage the day-to-day progression of an LPA Receivership in order to recover the bank's security.

The residual valuation method was the correct approach to take, and the general consensus with internal senior valuers, local agents and the auctioneer was that the property was worth £50,000 to £75,000. The fact that the property sold for well in excess of my valuation did not make my valuation approach incorrect, rather that there may have been a special purchaser or possibly a lack of redevelopment opportunities of that lot size.

The additional advertising generated a considerable level of interest, which we may not have obtained otherwise.

The strategy employed ensured that the security was recovered in 14 weeks from the date of our appointment. The strategy minimised holding costs and risk, while maximising return.

Throughout the instruction, I considered all relevant legislation, and took all necessary steps to protect the LPA Receivers' position, always considering their personal liability. Detailed file notes were made at all times regarding all telephone conversations.

I believe that I dealt with the instruction to the best of my ability and professional integrity. The lessons I learnt from this project are outlined as follows:

- The ability to communicate and negotiate effectively, verbally and through letters/reports, whether it is to the client, interested parties, solicitors, contractors or others.
- The ability to recognise the client's requirements and assess the situation, then formulate an appropriate strategy to achieve a result.
- Understand the need for a broad understanding of property issues. LPA Receiverships can present development or investment situations requiring landlord and tenant knowledge. Thus, I appreciate the need to involve surveyors from other disciplines, and objectively analyse their findings in order to provide informed and reasoned solutions.
- The importance in updating the client on a regular basis, and the adoption of a clear and concise method of reporting.
- To improve my time management and organisational skills in order to work to deadlines, and encourage other parties to do the same in order to achieve a common goal.
- To appreciate the need for lateral thinking and problem solving.

Appendices

The Appendices were Appendix I – Photographs, Appendix II – Site Plan, Appendix III – Notification of Appointment Letters, Appendix IV – Insurance Questionnaire, Appendix V – Reinstatement Cost Assessment, Appendix VI – Comparable Evidence, Appendix VII – Valuation Assumptions, Appendix VII – Marketing Details, Appendix IX – Silven Properties Case Law. Appendices II and VII are reproduced here.

Appendix III – Notification of Appointment Letters

Mortgagor

- Advising of our appointment.
- Notification that their powers to deal with the property are suspended.
- Advise that receivers now have rights to manage the property and receive rent.
- Advise that receivers have undertaken to provide insurance cover.
- Letter includes a copy of the Instrument of Appointment so that the mortgagor can receive legal advice if required, and independently validate the appointment.
- Informal information leaflet enclosed, advising of the day-to-day implications and the receiver's intentions.

Tenants (if applicable)

- Advising of our appointment.
- Requesting that all rents and other monies must be paid to the receivers.

Companies House

- Should also be separately advised by appointing bank by using Form 405(1).
- Confirming appointment as LPA Receivers NOT Administrative Receivers.

HM Customs and Excise

- Confirming appointment as LPA Receivers NOT Administrative Receivers.
- Requesting information on whether company has elected to exercise their option to tax in respect of the property.

(It is a criminal offence for anyone other than a licensed insolvency practitioner to act or purport to be an Administrative Receiver, so it is critical to ensure there is no confusion as to the capacity in which the Fixed Charge Receiver is acting.)

Solicitors

- Instructions to act on receiver's behalf – requesting confirmation of valid appointment and preparation of a brief report on title.

Insurers

- Advising of the appointment.
- Requesting that the property be placed on Receiver's Open Cover.

Statutory undertakers

- To gas, electricity, water suppliers.

Letter requesting opening of LPA Receiver's bank account

- Strict liability on the receiver to account for all monies received and distributed, and therefore necessary to maintain a dedicated bank account titled accordingly.
- Request overdraft facility to meet interim receivership costs pending disposal.

Appendix VII – Valuation assumptions

Valuation printouts from the software package used were also included in the report/appendices, with valuation assumptions set out below.

Appraisal 1 – Site clearance

- The property can be demolished at a cost of £7,500 including landfill tax etc.
- Discussions with the planners indicated that it would be acceptable to convert to flats within the existing structure, but considering site clearance and erection of townhouses, there would have to be an allowance for car-parking as the street is quite narrow with limited street parking. Therefore assume eight to nine townhouses.
- Construction will be over an 18-month period.
- Interest will be 2% above base – 5.75%.
- Each house can be sold for £60,000.
- Construction costs will be £50 per sq ft assuming very basic specification offered.
- Assume a profit on cost of 15%.
- Professional fees @ 10%.
- Contingency @ 3%.
- Stamp Duty 0% as site value is under £60,000.

Appraisal 2 – Conversion of existing building

- Gross Internal Area – 6,344 sq ft × 90% (allowance for common areas) = 5,710 sq ft – 8 flats × 713 sq ft/per flat = 5,710 sq ft.
- Each flat could be rented for circa £2,750 – £3,000 pa, and therefore it would be fair to assume a capital vale of c. £30,000 per flat.
- Conversion will be over a 12-month period.
- I discussed the scheme with our internal quantity surveyor and an outline costing of works was produced.

External Structural Repair	£10,000
Internal Rearrangement	£15,000
Basic decor and woodwork	£30,000
New windows and any other external works	£20,000
Plumbing and heating system	£15,000
8 bathrooms, 8 kitchens	£20,000
Electrics	£15,000
Total	£125,000

Thus a conversion cost of £125,000/6,344 sq ft = circa £20.00 per sq ft was applied.

- Profit on cost @ 15%.
- Professional fees @ 10%.
- Contingency @ 3%.
- Stamp Duty 0% as site value is under £60,000.

Conclusion

The appraisals demonstrate valuations given two different scenarios – subject to planning constraints, build costs and end sales.

The sensitivity analysis shows how sensitive the residual land value and profitability are to changes in build costs and end sales, and thus highlights how difficult it is to determine value accurately. Different parties in the market will have ways of saving on costs, and all will have different perceptions on any end values – thus making an accurate appraisal difficult.

The appraisals demonstrated that the property was worth more by retaining the existing structure for a flat conversion rather than site clearance and redevelopment. I adopted a value of £60,000 for the townhouses and given the character of the surrounding area, this was extremely optimistic but I did so to see where the parameters of land value lay within the appraisal.

The conversion appraisal indicated a land value in excess of £50,000, based on the assumption that it was possible to obtain eight flats. There still remained scope for more flats, and there is a single-storey extension which could possibly be extended further or raised to two stories. I was of the opinion that this would be an attractive lot in an auction room for local developers who deal in smaller lot sizes. The possibility of obtaining up to £75,000 with good attendance and competition in the auction room could not be discounted, particularly given current market conditions.

Illustration of issues/possible interview questions

Q Regarding the appointment of individual surveyors (ie your colleagues) as Joint LPA Receivers, are there any particular professional ethics/RICS issues here, compared with the more usual situation of a firm being appointed (eg client's accounts)?

Q Can you summarise the difference between the roles of LPA Receivers and Administrative Receivers?

Q What would the LPA receiver's position be in the event of the property owner's bankruptcy/liquidation?

Q What if there are assets within the property – has the LPA Receiver any power to sell these?

Q Can you explain in more detail why you considered it preferable to market the property without the benefit of an outline planning consent?

Q One reason you opted for auction was the ability to secure contracts – can you explain the auction process, and how contracts are established, applicants can see the contract in advance, etc.?

Q Had a special bidder been identified, such as adjoining owner looking, would auction still have been appropriate?

Q Can you explain how the joint agency basis worked?

Q Your instruction from the client – was this sole selling rights and what is the difference?

Q In respect of the marketing particulars, what information did you consider it prudent to include?

Q You referred to the Property Misdescriptions Act 1991 – can you give examples of specific aspects which could be sensitive to compliance?

Q I suspect that you received a number of low/speculative offers – do you have to pass all these onto the client?

Q Are there any other aspects which fall under the Estate Agents Act 1979 warranting consideration in the case?

Chapter 12

Telecoms Acquisition

This chapter examines the issues involved when telecommunications sites are acquired. The report is contributed by Sarah Kenney.

Introduction

During my time within the telecoms department in my employer's London office, I have gained experience of the telecommunications industry and been directly involved with the acquisition of radio base stations in central London. This had been predominantly for (Tel-com) who are a Licensed Telecommunications Operator.

My critical analysis will examine my role in the acquisition of a telecoms base station for my client, Tel-com, which I worked on during the second year of my training period for the APC. I have chosen this case for my critical analysis as I feel it will demonstrate the skills and experience I have obtained in my specialist area of surveying during my probationary period.

Glossary of terms

This report contains a number of technical terms, which relate to the tele-communications industry; therefore an explanation of these terms is included in Appendix I (not reproduced here).

Outline of critical analysis

The report is divided up into sections which relate to the distinctive stages of the acquisition, with a final chapter focusing on the important lessons I have learnt through my participation in this case.

I will explain the process from my client's initial instruction for the survey and my recommendations to my client, through to the design visit. I will then describe the negotiations with the landlord, submission of the planning application and the legal completion of the acquisition. This will be followed by a reflective analysis of the case, and my role.

Instruction

Agreeing instruction

Upon receiving instructions from my client to acquire a telecommunications base station, I confirmed the fee basis for the work in line with our main contract of engagement, which also sets out the scope of work required by my client.

Conflict of interest

Prior to commencing work on this project, a conflict of interests check was carried out in line with the RICS Rules of Conduct and my employer's internal procedures. It was established that there were no conflicts.

Understanding the client's requirement

After confirming the instruction, I meet with my client to establish the specific requirements in relation to this acquisition, and their time scale for this project.

The radio planner for Tel-com had identified an area around Arthur Street where they have a deficiency in coverage, and therefore required a new telecoms macro cell (as shown in Appendix II – not reproduced here). My client's requirement was to acquire a new cell in the search area within six months.

Client's objective

My client's objective with this instruction was to improve their 2G coverage in this area to ensure there is in-building reception for their clients. It also forms part of their 3G network rollout where they have a licence requirement to provide coverage, and as such was an important site.

Confidentiality

Prior to writing this report, I obtained my client's consent to use details relating to the acquisition of a telecommunications base station at 166 Arthur Street in London. In the interest of confidentiality, my client requested that the agreed rental figure be omitted from the details.

Site survey and recommendations

Preliminary enquires of search area

My first task was to identify premises within the search area which would be suitable for my client and meet their requirements – which are buildings of a similar height or taller than surrounding buildings so the signal from the antenna would not be interrupted. My client also requires a building where planning sensitivities are avoided. The final point is that there is willing landlord with whom a lease could be negotiated.

Arthur Street is a commercial area located in the City of London. The surrounding area consists predominantly of office buildings with very few open spaces. I was therefore aware I would be looking for a rooftop site.

In order to speed up the process and meet my client's time scales prior to undertaking a physical search of the area, I carried out a desktop survey which ascertained the number of properties that would be available in the area. I considered possible sites which had been previously drawn to my attention by landlords, because if I could identify a suitable building, it would have help the acquisition time scale. Unfortunately there were no such buildings within the search area.

I also made initial enquiries with the local planning authority, which is the City of London, in order to establish any special land designations for the area, and to confirm that this site was within the Arthur Street conservation area. The significance of this was that my client could not use their permitted development rights for an installation, and would therefore be required to submit a full planning application. This had the potential of prolonging the acquisition time scale.

I also spoke with the duty planning officer to find out if there were any listed buildings or other properties in the area where they would rather not have telecommunications equipment installed. This was in order that I could carry out an informed survey, and not pursue options which would result in planning objections which would delay the process.

Walk around site survey

After my preliminary checks, I carried out a walk around survey with the radio planner who identified buildings which appeared suitable, and which were felt would meet technical requirements. While out on site, I noted details of the tenants and managing agents for the buildings identified by the radio planner, and recorded their location and addresses on a map for reference (see Appendix II – not reproduced here).

Follow up

Once back in the office, I contacted the managing agents for the buildings which had been selected on the survey, and established if the landlord was, in principle, interested in negotiating lease terms.

Where there had been an expression of interest, I followed up my initial contact with a letter outlining my client's requirements in more detail. A copy is included in Appendix III (not reproduced here). I included my client's standard Head of Terms, in order that the landlord or agent would understand from the outset the terms my client would be seeking to agree.

By providing this information prior to proceeding with the acquisition, it ensures that parties who have not dealt with telecoms matters in the past could gain an understanding of the parameters in which my client operates. It means that landlords can make a better informed decision on whether to proceed, thus ensuring the acquisition process runs more smoothly and keeps within my client's time scale requirements.

List of options

After establishing which landlords were interested, I compiled a map for my client which detailed the buildings where there was a willing site provider – a copy of which is in Appendix IV (not reproduced here). The radio planner then confirmed his favoured option and requested an initial roof top survey.

Roof top survey

I liaised with the agent or on site building managers regarding the options favoured by the radio planner to access the roof. The radio planner needed a roof top survey to confirm that surrounding buildings and clutter would not interrupt the antennas. It also allowed me to establish whether a scheme could be achieved which was sympathetic to planning issues, and confirm there was sufficient space on the roof for my client's equipment cabinets.

Traffic Light Rating

Following the roof surveys, I was able assign each property a rating in accordance with the 'Traffic Light Rating' model, as shown in Appendix V (not reproduced here).

This determined the most appropriate community consultation strategy, based on the model adopted by the mobile operators – a copy of which is in Appendix VI (not reproduced here). This is in line with commitment 1 of the operators' Ten Commitments – as shown in Appendix VII (not reproduced here). The Traffic Light Rating also helped me advise my client as to which sites were sensitive in terms of planning and community issues, and which could lead to problems and delays during the acquisition process. The properties favoured by the radio planner were all rated 'green'.

Site specific option report – recommendation to client

The roof surveys highlighted that only two of the initial six properties would be suitable. Based on my preliminary discussion with the local planning officer, I was aware other operators had gained planning permission to install equipment on 166 Arthur Street, and therefore a precedent had been established for the building. I also knew from my initial approach to the landlord that they were interested in negotiating a lease, as they had agreed terms with the other operators. Therefore I recommend to my client in the form of a 'Site Finders Report' that the most suitable option to pursue was Arthur Street. A copy my report is in Appendix VIII (not reproduced here).

I did, however, maintain contact with the agent handling the other properties in case the acquisition of 166 Arthur Street failed at any stage. In this case, the second property could be used as a back-up option, ensuring that there would not be lengthy delays to my client's time scales.

Nomination

Arranging design survey

Following my recommendation, my client nominated 166 Arthur Street as their option for this cell, and appointed consultants to design a scheme for the installation of their equipment on the building.

In the meantime I contacted the agent and confirmed that my client was interested, and also arranged a visit for my client's contractors to carry out a detailed design survey. I also asked for initial feedback on the Heads of Terms that I had previously sent the agent, and he confirmed he would respond when the drawings had been produced.

At this time, the agent informed me that two other mobile phone operators were in discussions with them, and he provided me with their drawings so I could ensure my client's scheme did not conflict.

Detailed design survey

At the design visit I liased with the on-site building manager to identify the locations on the roof which were available to my client. I advised the design consultants on the scheme to ensure that the plan produced would be suitable from a town planning perspective.

Acquisition

Compiling comparable evidence

While I waited for the scheme drawings to be produced, I spoke with the acquisition agents for the other operators in order to gather evidence of the terms and rent they were agreeing with their respective landlords. I also compiled comparables of other acquisitions that had recently been completed by my client in the area.

Scheme drawings

Once I received the scheme drawings from the designer I reviewed them to ensure they were accurate and reflected the proposal that had been agreed on site. A copy of the drawings are in Appendix VI (not reproduced here).

Pre-consultation with local planning authority

When the drawings had been approved, I sent a copy to the local planning authority for consultation. A copy of the letter is in Appendix X (not reproduced here). The purpose of the consultation letter was to invite the planning officers to provide feedback on the scheme prior to a full planning application being submitted. This was in line with commitment number two of the mobile operators' Ten Commitments (as shown in Appendix VII).

After a couple of weeks I followed up the letter with a call to the planning officer who informed me he had no specific comments on the scheme, and the fact that two other operators had secured planning consent for the site would act in my client's favour.

Negotiations with landlord

I sent the design drawings to the landlord's agent for approval, along with a copy of the Heads of Terms which I had revised since my initial approach. This was based on the information compiled from my client, and the agents for the other operators looking at the building. The agent confirmed that the scheme was acceptable to his client, but with amendments to the Heads of Terms. While negotiations continued, I sought consent from the agent to submit a full planning application.

The agent had sought to include a break provision for the landlord after the third year, on six months' notice. My client was willing to accept a guaranteed minimum term of three years, but I wanted to qualify the basis on which the break could be actioned. I felt that the landlord wanted the break provision to safeguard any future redevelopment proposals, and so added this as a condition of the landlord's break. I also extended the notice period to 12 months in order to ensure my client would have sufficient time to acquire a replacement site if their lease was terminated.

My client instructed me to seek consent to an 'Early Access' agreement which would allow them to begin work and install their equipment on the site prior to the lease completing. However, the agent confirmed that his client required full legal completion before any works commenced on site.

The landlord's agent accepted my proposed revisions of the terms outstanding, and confirmed that we had a set of Head of Terms which were in an agreed form. These can be seen in Appendix XI (not reproduced here).

Throughout the negotiations, I continued to report progress to my client on a regular basis, and sought their instruction on any points which deviated from their standard Heads of Terms.

Co-location approval

As there were other mobile phone operators with schemes already approved by the landlord, I was required to send out my client's drawings to them for co-location approval. This is an industry established procedure, and a copy of the form submitted is in Appendix XII (not reproduced here).

Planning

Obtaining planning permission

Once a suitable property had been identified, I needed to obtain planning permission prior to being able to conclude the transaction. When conducting my initial planning enquires, I had identified that the property was within a conservation area, and that I needed to consider how this would affect the type of application which could be submitted to the local authority.

Key planning issues

With regards to installations carried out by licenced telecoms code systems operators, there are three types of applications which can be submitted to the local planning authority. These are a Licence Notification, a GPDO application or a Full Planning application.

The main piece of Town Planning legislation which relates to telecommunications installations is Part 24 of the Town and Country Planning (General Permitted Development) (Amendment) (England) Order 2001 – Development by Telecommunications Code System Operators.

If the proposed installation is of a minor nature and does not exceed the criteria in Part A.2(4), then the works can be dealt with by way of a Licence Notification submitted to the LPA. This is a requirement of the 1984 Telecommunications Act.

If the proposal is permitted, as defined by Part 24 Town and Country Planning (General Permitted Development) (Amendment) (England) Order 2001, but exceeds the criteria in Part A.2(4), a GPDO application needs to be made.

However, if the proposed apparatus is not permitted, a full planning application will be required. I had previously established that 166 Arthur Street was within a conservation area, and in being article 1(5) land, would not be permitted development under Part 24 of the GPDO. Therefore a full planning application would need to be submitted to the local planning authority. I had already advised my client of this fact when I forwarded them my Site Finders Report.

Submission of full planning application

While I continued to negotiate with the landlord's agent, I prepared and submitted a Full Planning Application for the proposed installation. I decided that it would be beneficial to start the planning process while continuing negotiations, rather than wait until they were completed, to ensure the acquisition process was not delayed.

From my consultation with the planning officer, I was aware that they were, in principle, happy with the scheme. I compiled material to support my client's application, and also drew on national and local planning polices, such as PPG8 to back-up my client's scheme. A copy of the full planning application is in Appendix XIII (not reproduced here).

While the planning application was with the local planning authority for determination, I liased periodically with the case holding officer. The purpose of my contact with the officer was to ensure that if he received any negative responses during the consultation period, I could deal with them immediately. This meant that if any amendments were required to ensure planning permission would be granted, these could be undertaken while the application was current, rather than wait for a possible refusal before responding.

In this case, the planning application passed smoothly through the town planning process and the planning officer granted planning permission through delegated powers. A copy of the planning consent can be found in Appendix XIV (not reproduced here).

Acquisition completed

Solicitors instructed

After planning permission had been obtained and the negotiations with the landlord's agent successfully completed, I prepared instruction documents for my client's solicitor.

Once a case holding solicitor had been appointed, I kept in contact with them while the wording and form of the lease was being agreed.

File closed

On completion of the lease, I copied the salient parts of my case file and sent this to my client for their records. I then agreed and submitted my invoice, in line with the previously agreed fee basis for the work undertaken.

Reflective analysis

Analysis of the case

This case has allowed me to follow an acquisition through from instruction to completion. Although the site acquired by my client is small in scale, I feel that many of the principles and considerations are the same as would be encountered in transactions for larger leaseholds. The landlord for 166 Arthur Street is a commercially minded owner, and his agent's approach to negotiations were the same as if it had been for a different property type.

Experience gained

This case has allowed me to improve my professional ability, and during the course of the acquisition I have developed my skills in the area of General Practice. I feel that many of the skills I have developed as part of this acquisition can be transferred to other projects.

This project has highlighted the importance of close liaison with all the parities involved. For example, through my continued contact with the local planning officer, I ensured that planning permission was successfully granted to my client. From the start of the project I had to interact with a host of individual people with different roles – such as my acquisition manager, my client's radio planner, build and design consultants, planning officers and other agents.

I have learnt the importance of keeping clear concise file notes following telephone conversation to ensure I comply with internal Quality Assurance Standards, and do not lose track of events and negotiations.

This case has allowed me to improve and develop my skills in both letter and report writing. I have learnt the necessity for attention to detail and accuracy when corresponding with planners, other agents or solicitors, and when reporting advice to clients.

I have learnt the importance of time scales and how preparation (such as carrying out desktop enquires before going out on survey) helps ensure that matters progress smoothly. I have also seen how compiling comparable evidence prior to commencing negotiations can help strengthen your position, and ensure you deliver the best deal to your client.

Through the preparation of Heads of Terms and negotiations with the landlord's agent, I have deepened my knowledge and practical experience of both the Landlord and Tenant Act 1954 and the Telecommunications Act 1984.

I have also developed skills with regards to negotiating and have learnt the importance of identifying the areas important to my client to ensure these are not concede and are included in the lease. On reflection, it may have been quicker to meet with the landlord's agent to agree terms rather than rely on the exchange of correspondence and telephone calls as this would have meant that all the terms could have been discussed and negotiated at the same time. I can carry this lesson forward to future cases.

I have gained a better knowledge and understanding of the planning system, which will be beneficial in future cases. I now have the confidence to liase with planning officers to ensure a successful application. I feel I have also developed an understanding of the information required by a planning officer to enable them to determine an application and the importance of supplying supporting material to strengthen my client's case for consent.

Appendices

The Appendices were: Appendix I – Glossary of Terms, Appendix II – Map Showing the Search Area Defined by Tel-com, Appendix III – Copy of a Letter Sent to Potential Site Providers, Appendix IV – Map Showing Properties with an Interested Landlord, Appendix V – Traffic Light Rating Model, Appendix VI – Consultation Strategy, Appendix VII – Ten Commitments, Appendix VIII – Site Finders Report, Appendix IX – Design Drawings, Appendix X – Copy of Pre-Consultation Letter, Appendix XI – Draft Heads of Terms, Appendix XII – Co-location Form Sent to Other Operators, Appendix XIII – Copy of Full Planning Application, and Appendix XIV – Copy of Planning Consent Letter. None of the Appendices are reproduced here.

Illustration of issues/possible interview questions

Q You referred to no conflicts existing. Can you give an example of how a conflict could have been present?

Q Does your employer also act for landlords, and, if so, would this create any conflicts?

Q What exactly is the role of the radio planner?

Q What technical constraints are there regarding telecoms sites in built up areas?

Q Out of interest, how do you protect a tenant client against a new development taking place which blocks out the signal to the telecoms facility?

Q Did you consider the option of 'twin tracking' in respect of the planning application, and what does this mean?

Q Can you explain the traffic light test?

Q Can you please explain the importance of pre-consultation with respect to the operators' 'Ten Commitments'?

Q You did not mention that the lease was contracted out, but I understand that telecoms leases often are. What does this mean?

Q The Telecommunications Act 1984 provides 'Code Powers'. What are these, and what would be the implications in this particular case?

Q Did you need to restrict the landlord's ground for exercising the break to development, and if so, why?

Q Regarding a possible Early Access agreement, how would you advise a landlord under similar circumstances to those in this case?

Q What are the principle terms of Early Access agreement, and does this commit the landlord and completing a lease?

Q Let's now assume that you had been asked to undertake an investment/capital valuation of the facility, with the rent/lease/tenant in place. Where would the risks lie to an investor, and what valuation methodology would you adopt?

Chapter 13

Marketing – Further Issues

This chapter provides supplemental information in respect of marketing/agency. The material is provided by Midlands Property Training Centre, based on training initiatives facilitated for general practice graduates.

The essence of these initial notes is that the marketing of property involves more than advertising and promotional initiatives. They demonstrate why best value can be secured on behalf of an employer or client only when the opportunities for a property have been effectively formulated prior to marketing.

The objective of marketing will be to secure best value on behalf of an employer or client. This will typically mean aiming to achieve the best price. However, other objectives could include the need for certainty that a transaction will take place, and the timing of the transaction. Budgeting/financial issues may be relevant.

Relevant factors

There are numerous factors which may be considered prior to the disposal or letting of property interests. Examples include:

Considerations of alternative uses

- Alternative uses may be more valuable than the existing use.
- The possible effect of on-stream developments on supply and demand may need to be considered.
- It may be necessary to commission specialist research.
- Development/investment appraisals may have to be undertaken.
- It may be beneficial to acquire neighbouring interests.
- Refurbishment may be a possibility.

Planning issues

- It may be beneficial to secure planning consents before exposing the property to the market.
- Alternatively, a development opportunity could be advertised with a view to working with a developer to secure consent.
- The marketing material could contain a letter from the planning authority outlining their informal views.
- An element of 'hope value' for alternative uses can enhance a sale price.
- If the planners are likely to receive enquiries, they should be alerted prior to marketing. If given advanced notification, their consideration of the opportunities for the site may be of a more positive nature than if they are suddenly required to respond, and forced to give a inevitably cautious opinion.

Opportunities to improve investment value

- A tenant may need to be found prior to advertising an investment property for sale – the capital value of an income stream being substantially greater than the capital value based on vacant possession. This is more likely to be the case with a good quality, well located, investment property.
- In other situations, the value of the property for owner occupational use may be substantially greater than the value of the property as an investment. This is more likely to be the case with properties in secondary locations.
- It may be beneficial to conclude a rent review before selling the property – again because of the impact on investment value.
- Other opportunities to optimise the investment profile of a property include the grant of a sufficiently long lease and the securing of a sufficiently strong covenant.
- The granting of a rent-free period or other incentives may allow a headline rent situation to be constructed.
- It may be preferable to secure an optimum mix of uses – such as within a retail development.
- A client's neighbouring interests could present opportunities, but also limitations.

Timing

- It may be preferable to wait until market conditions have improved before selling the property.
- It may be possible to secure short term income in the meantime.
- Marketing could be on the basis of 'to-let or for sale'.

Lease variations

- It may be possible to make an interest more attractive by negotiating lease variations with the landlord – such as where a lease is to be assigned, or where a subtenant is being sought.
- A subletting may be preferable to assignment.
- In some situations, occupiers may be more attracted to a direct relationship with the landlord than to a subtenancy arrangement.

Reducing uncertainties

- Purchasers/investors often discount disproportionately highly for risks and uncertainties.
- A cautious view may, for example, have to be taken on the likely level of rent that could be secured at rent review, or the time that will be taken to find a tenant on reletting.
- In the case of development sites, a large contingency may have to be allowed for the uncertainty in respect of the cost of clearing contamination. There may uncertainty as to the intensity of development that will be capable of securing

planning consents – as well as uncertainty in respect of associated delays and costs.
- In other situations, those uncertainties will actually present opportunities for prospective applicants. For example, the ability to take a particularly bullish approach to an outstanding rent review may be attractive to certain investors.
- Some developers may be able to add value because of their special expertise in certain areas.
- There could be many means by which individual bidders can create special opportunities for themselves.

Client requirements

- Clients may have specific objectives or requirements relating to operational aspects of their business. Sometimes, decisions taken for business reasons may mean that, in property terms, the most commercially preferable course of action is not feasible.
- Clients/owners may also, of course, have specific requirements in connection with the property, and/or the way it is marketed.

Making it easy for applicants

- The research time and the potentially abortive costs incurred by applicants should be minimised.
- Information established by the surveyor prior to marketing could include: availability of services, rateable values/appeals/rates liabilities, service charge arrangements including copies of accounts, contamination surveys or structural surveys, title problems, lease restrictions and planning information
- Additional information should be readily available.
- Clients should be in a position to respond quickly to recommendations made by the surveyor in respect of any offers received.
- Solicitors should be able to submit documentation as soon as terms are agreed.
- All parties should be able to provide swift and complete responses to additional queries.

Missing the market

- Marketing may not be effective when the attractiveness of an opportunity is exaggerated.
- The parties looking for the type of opportunity that has been so attractively described may, on viewing, be disappointed that it falls short of their expectations and business requirements, and thus not pursue their interest.
- At the same time, the parties suited to such low cost/low quality accommodation may perceive from the promotional headlines that the property is of a higher quality (and therefore higher cost) than is necessary for their requirements, and not pursue matters further.

- The advertising therefore misses its market. 'Telling it how it is' may be the most effective approach.
- For larger, higher profile properties however, decisions made by occupiers and investors can be significantly influenced by the quality of promotional material.

Pricing tactics

- A good illustration of how psychological factors can affect property transactions is the way buyers of residential property sometimes attribute more weight to the reduction from an asking price, than to the actual purchase price.
- A relatively high figure can be quoted to provide room for negotiation, but prospective applicants may be more attracted to alternative, more realistically priced opportunities. Applicants generally wish to minimise the time spent searching for suitable premises, and may not devote resources to establishing the extent to which terms may be negotiable.
- Pricing tactics should always reflect market conditions and the expected strength of interest.
- Some of the phrases commonly used include: 'terms negotiable', 'concessions available' (these can demonstrate flexibility, and realistic price aspirations), 'offers invited', 'sale by mortgagee in possession' (the property would seem to be there to be sold, and there may be the chance of a bargain), 'in the region of', 'in excess of' (this helps guide applicants – as does an indicative price range), 'expressions of interest invited' (this could be used where it is inappropriate to quote a figure such as where a meaningful valuation cannot be undertaken), and 'price available on application' (this may be necessary where price publication is sensitive).
- When surveyors are unwilling, or appear unable, to give an estimation of the price they expect to achieve, applicants can become frustrated. Price indications can be given verbally if need be. However, with tender exercises, for example, care has to be taken not to mislead, or provide inconsistent information to applicants.
- The provision of over-cautious opinions of value can sometimes reduce prospective bidders' opinion of value. It should be noted that asking rents can indicate what was in the mind of the valuer and, particularly where comparable evidence is weak or non-existent, can sometimes influence rent review, lease renewal or other settlements to the detriment of the owner/ client.

Marketing/letting illustration

A brief illustration is shown below of some of the issues arising when considering the most suitable advertising outlets when letting property.

It is assumed that the property is vacant and to let – and the landlord requires a new tenant.

The surveyor will be involved in aspects such as property inspection, rental valuation, etc, but in relation to the search for a tenant, the following issues will be of particular relevance:

The target market for the property

- What sort of tenants occupy such accommodation?
- Who are surrounding occupiers?

Marketing material/advertising outlets

- Erection of 'to let' boards.
- Preparation of marketing particulars.
- Is an *Estates Gazette* advertisement appropriate?
- Is a local newspaper advert appropriate?
- Are there any other property, business, and/or specific trade media outlets?
- Are there any known requirements in the market?
- Mailshots to agents.
- Mailshots/approaching the landlord's other tenants (if any)/approaching other occupiers in the area.
- Any other opportunities – websites, etc.
- Are there any special features/marketing attractions relating to the property that can be identified?

Marketing budget

- What is normally recommended?
- What are the client's views (realistic or unrealistic).
- What if the property does not initially let?

Lease terms, incentives

- What are the proposed lease terms, will they be sufficiently attractive to prospective tenants (balancing the landlord's wider investment objectives/ high capital value and a tenant's need for flexibility – see below), and how negotiable are the proposed terms?
- Can any incentives be granted – such as a rent-free period?

Enhancing capital/investment value

- How can a letting benefit the capital value of the property as well as providing a rental income. Capital value will be helped by finding a tenant of good financial standing (ie of good 'covenant strength') and by a long lease length. This is because landlord investors like security of income (provided by a tenant who is unlikely to go bust and who is committed to many years of rent payment). Such factors also improve the ability to raise finance, noting the importance of this to investors.

The method of disposal – sales

A method of disposal should enable the widest and most appropriate audience to be targeted, and the best deal to be secured.

This could be auction, private treaty or tender (informal tender or formal tender). An auction sale is typically where the property is included in auction alongside other properties, although an auction could, in fact, take place for an individual property. Private treaty involves prospective purchasers being able to negotiate with the owner/owner's agent and conclude a deal. Tender is where offers are made by prospective purchaser's by a given date, and then considered by the owner/owner's agent. Formal tender is where the offer can be accepted by the owner, and the purchaser is therefore contractually committed. With informal tender, the bidder is not contractually committed by making an offer, and as with private treaty, the parties are contractually committed on 'exchange of contracts'.

Some of the basic features of auction, private treaty and tender are as follows:

Auction

- Open bidding.
- Bidders are aware of other offers (and possibly the identity of other bidders).
- Identity of bidder cannot be controlled.
- Commitment to buy/sell on the fall of the hammer.
- Guide price can be a cautious estimate of value/can capture interest.
- Reserve price is the minimum price acceptable.
- Contract terms and the timing of completion are fixed.
- Tactical delay/renegotiation of terms is avoided.
- Purchasers must be in a position to complete/funding needs to be available.
- Uncertainties and the prospect of abortive costs may deter prospective purchasers.
- Best bidder only has to marginally outbid the second highest bidder.
- In weak market conditions, auction may be seen as an outlet for distress sellers.

Private treaty

- A perception of exclusivity can be created for prospective bidders.
- This may be particularly beneficial in weak markets or for properties of special interest.
- Private treaty negotiations could still lead to competitive bidding/the conducting of a tender.
- There is scope to consider all uncertainties and opportunities.
- There is scope for the parties to work together.

Tender

- Parties are bidding blind to each other's offers.
- This may be a better means than auction to extract the best price from a special bidder.

The method of disposal considered most suitable will reflect a range of factors, including the characteristics of the property, the nature of its market, and economic and property market conditions generally.

Study tasks

The following can be considered by individual private study, or by group discussion.

For the marketing situations indicated below:

- Consider whether any opportunities can be undertaken prior to marketing that could help enhance value and/or help meet other requirements.
- Consider what method of disposal might be appropriate, and why.
- Consider what sort of marketing strategy could be adopted.

Where further information may be required in practice, this should present further issues to consider.

- 3,000 sq ft/300 m² unit in industrial estate near city centre. Tenant required.
- 1,000 sq ft/100 m² office suite in recently constructed office block. Good city centre location. Tenant required.
- 1,000 sq ft/100 m² retail unit in shopping centre. Tenant required.
- Lease available for assignment. High Street retail premises in a market town, let to an independent/local trader. Six years to expiry.
- High Street retail investment property let to a national retailer, having 10 years to lease expiry. Disposal of freehold required.
- Builders yard/open storage facility. Tenant required.
- Haulage yard. Held on lease with two years unexpired. Residential development possibility. Maximum potential to be realised on behalf of an owner who has no urgency to sell.
- 200,000 sq ft/20,000 m² warehouse/distribution centre located by motorway junction. Completion of construction imminent. Built speculatively. Developer requires sale ultimately, but can retain the property in the shorter term if need be. Investment value/sale price to be maximised.
- 50,000 sq ft/5,000 m² 1960s office building in provincial city centre. Half let. Market rental values at a third of modern accommodation. Owner wishes to sell but will consider all options.
- Poor quality, small retail premises in run down residential area. Recently repossessed by bank who require sale. Reasonable economic outlook. Similar properties in the area have appealed to both local traders and local landlords/ investors.
- Four plot residential development opportunity. Good location. Outline planning permission granted. No planning issues to overcome. No development/construction uncertainties. Reasonably buoyant market conditions. Sale required.
- 20 acre greenfield site suited to residential development. No planning consent secured. Planning problems expected. Planning gain requirements likely.

- Site suitable for B1 development. Extent of contamination unknown. Possible problems with migration onto adjoining site. Disposal required.
- Four-bedroom house in good quality residential area. Sale required.
- Residential investment property in student area. Works need doing. Sale required.
- Client in financial difficulty. Needs to dispose of hotel as soon as possible. Town centre location. Good levels of trade.
- Local authority wishes to dispose of around 50 properties, comprising retail premises and industrial units let to local businesses.

The above concentrates on the position of the owner/investor seeking to dispose of their interest. In examining the issues again, consider all the issues from the perspective of the purchaser or tenant requiring property. This should also help consider how purchasers can best be attracted.

Wider factors associated with marketing

There are wider factors that can affect the success of marketing initiatives.

- Within your organisation, is marketing literature available to applicants without them having to speak to the case surveyor. They may not wish to spend time discussing their requirements when merely requesting a set of details. Also, what happens if the case surveyor is absent?
- After an instruction has been won, who is the point of contact for clients? It may not be best practice for junior surveyors to become the main point of contact after more senior surveyors have earned the instruction.
- Do receptionists, support staff and all surveyors regard themselves as part of the sales pitch for any properties being marketed, and also for the company?
- Can you think of any other opportunities within the company by which marketing exercises can be undertaken more efficiently?

Chapter 14

Office Development

This chapter examines the issues involved in the development of office accommodation.

The report is contributed by Honor Boyd.

Introduction

The subject of my critical analysis is the development of a block of offices in an old mining town in Leicestershire. The site was purchased with a derelict property located on it, and had been on the market for a number of months prior to the purchase.

The critical analysis demonstrates the decision making process and actions that must be taken when developing a property with joint funding responsibility. It also demonstrates the process undertaken when obtaining planning permission for a property within a conservation area.

The reason why I have chosen this particular project is because I was given a great deal of responsibility for the development, and also because it best reflects the areas of surveying upon which I have concentrated over recent years. The development of these offices was very interesting, if not a little challenging at times.

This submission does not contravene any confidentiality requirements of the council or the main funding body (local regional development agency). I have, however, been asked to keep financial information and adjoining site owners' names confidential.

Client brief and key issues arising

I became involved in this development just as the use of the site was being considered. This followed a demand study which had been carried out for the council.

My role was to obtain planning permission for a development that would serve the local community and also new or embryonic businesses. I was also given the responsibility of overseeing the development once the professional team and contractor had been appointed. The RDA was the main funding body of the project and as a result carried a great deal of weight with regard to decisions that were to be taken.

- The importance of due diligence, including the assessment of the client's requirements.
- Planning and development methodology, including assessing the content and fit of the development plan.

- The economic strategy for the area, and how the proposals for the site would fit.
- The need for adequate funding.
- The need to resolve a number of legal implications arising.

Due diligence

As with any activity, it is essential to follow a procedure of due diligence prior and during any activity. This ensures that potential problems may be highlighted at an early stage, and all options are considered. The due diligence process is designed to ensure that a surveyor is in possession of all the facts prior to carrying out any activity. I felt that I needed to be aware of the following:

- The characteristics of the local market.
- The condition of the surrounding land.
- Any policies contained within the Development Plan, Statutory Instruments, Circulars and Guidance Notes that may affect a proposed development.
- Any recent developments that may have set a precedent.

Planning and development methodology

Obtaining planning permission was my main role at the early stages of the development. It was essential that I maintained close contact with the planning department in the council. This was imperative given that the subject property is within a Conservation Area, and as most planning authorities are looking for high quality design of a specific nature in such areas.

I maintained close contact with the Conservation Officer, who provided me with a great deal of advice regarding the type of materials and the style of property that would be granted planning permission. Many older properties within the town were constructed from a local jumbo brick, which was historically designed in order to avoid a 'brick tax'. Initially the Conservation Officer had suggested that we should have replica jumbo bricks made for the development. Had he enforced this suggestion, the building would have been too costly.

A second difficulty that I encountered was that the property did not have the space to provide the required parking spaces. However, had the existing property been refurbished, there would have been no parking spaces at all as there was no vehicular access from the road on to the site. In order that the planning permission would be granted without supplying the required level of car parking I highlighted to the planning department that there was substantial car parking provided throughout the town.

When the project was first discussed it was necessary, as with all potential development schemes, to identify the site within the relevant Local Plan (The Council Local Plan Deposit Draft February 1995). The site may well be already identified for an alternative use, and as a result it would not be possible to obtain planning permission for the particular development which may be suggested.

When the local plan was consulted, it was noted that the site was within the town's Conservation Area and as a result was affected by policies E10 and E11 (see Appendix F – not reproduced here).

Economic strategy/economic development

As this project was carried out by the Regeneration Division of the council and a great deal of the necessary funding was provided by government bodies, the relevant economic strategies must be adhered to. The mission statement for the Regeneration Division is 'to improve the quality of life of all the people in Leicestershire by promoting equality, democracy, community welfare and sustainable economic, social and environmental well-being through the provision of best value services'. Part of this statement involves support and encouragement for local businesses, and the provision of suitable accommodation for small businesses, in order that an area may be economically regenerated.

Part of the strategy of the division involves a programme of land reclamation and property development, in order to improve the character of many parts of the District which were previously blighted by derelict land and the decline of the mining industry. Substantial investment was secured from central government and the European Union for the reclamation of two collieries within the area. This has been successfully completed, and now the Council is in the process of developing a number of commercial properties – the new offices in the town being one of these developments.

Funding

Prior to carrying out any work on site, it is essential to secure the necessary funding. Most of the funding was secured by my colleague, with the majority of funds being provided by the regional development agency. As part of the funding arrangement a Development Agreement was drawn up. Given the fact that a Development Agreement is a legally binding document, it is necessary to consult and follow this agreement at all stages along the way, as deviation may render the funding invalid.

Once all the necessary funding was secured and in place, I monitored all the relevant budgets and completed the necessary grant claim forms in order that the money could be drawn down from the relevant body. In order that I could do this effectively, I set up a spread sheet which enabled any increase in costs or fees to be monitored, and any problems addressed as soon as was evident. While we had agreed all costs with the professional team prior to going on site, there were unforeseen difficulties that arose once on site that lead to certain increases in some of the fees.

Given that the relevant property was within a Conservation Area, it is within the rights of a Planning Authority to check that the necessary funding is in place prior to granting planning permission, in order that a high standard of development may be guaranteed. If this precaution was not used, it may be possible for a developer to start a development and subsequently find a deficit in finances, and as a result reduce the specification of the property.

Legal implications

It was necessary to ensure that the subject property was within the ownership of the developer, and also to ensure that the boundaries were in place, as failure to do this may result in certain legal implications. It is also necessary to look at things such as drainage in detail (prior to starting a development), as this may result in legal implications (for example, it may be discovered that it is necessary to put an easement in place prior to the property being fully serviced). If essential checks have not been made, then a development can be held up while the necessary legal procedure is followed.

When drainage was considered for the property by the engineer, there were a number of options, one of which involved pumping the drainage to road level. This, however, would have proven too costly and as a result an alternative option needed to be considered. This alternative was to drain into the adjacent builder's yard. When I considered this option in detail, I was concerned at the fact we would need to put an easement in place, as in many cases this can prove to be costly, and we were short of time at this particular point. We went on site and discussed the possible cost implications with the adjacent site owners' agent, and he advised of the figure his client would be prepared to accept in order to put the easement in place. This cost was also to include both sets of legal fees. After much deliberation about this additional cost I looked into the history of the site at the suggestion of my supervisor at that time. As a result, I established where the water had drained prior to the demolition of the previous building. I discovered that the water from the previous building had drained into the adjacent site. This in effect gave us existing rights and as a result an easement was deemed unnecessary.

Development

Planning permission was obtained with a number of conditions attached, such as the type of materials that must be used. There were also a number of other alterations made to the original plan, which have added greatly to the cost.

The development reached practical completion at the end of August 2001. The development now provides a total of approximately 276 m^2 (2,874 sq ft) over three floors, and 13 car-parking spaces. Each floor will have a central core area providing a kitchenette and toilet. There has already been substantial interest shown in these units with a couple of active inquiries, prior to any marketing. The marketing and management of the property has now been taken over by the Estates Office within the council.

Critical appraisal

One issue that I had to consider very closely was the fact that the property was contained within a Conservation Area. This can add great cost to a project that is often not accounted for. When I tried to reduce the cost of the project in a number of areas as a result of increasing costs, I discovered that we were restricted as to the type of the material that was insisted upon, and these materials could not be changed for a lesser standard. While I had built a contingency sum into my overall figures, the additional cost was far greater than the contingency allowed for. It is essential that all issues are both known and in turn considered prior to the start of any project.

There were many different issues that arose at each stage of the project:

- One of the issues that faced me in this project was the fact that the planning department was more concerned with the historic fabric of the town, as opposed to the property development needed to underpin the commercial activity necessary to regenerate the town.
- Another issue which I had to deal with was contamination. Initially there were a few surveys carried out prior to the demolition of the existing property. However, it was impossible to assess the condition of the ground underneath the property until after demolition. As a result, alongside my supervisor, who specialises in contamination, a number of surveys were arranged. As a result of the previous surveys, we were of the opinion that remediation work would not be necessary and that while the top few layers of the ground were man-made, the remainder was natural and therefore able to support the proposed development. However, this was not the case and the man-made ground ran a great deal deeper than what we originally anticipated, and as a result we had to deep pile the foundations (something that we had not prepared for in our funding). As a result of this, additional funding was necessary to complete the project. Given the fact that records are often incomplete in areas such as the subject town, it is very difficult to predict such an occurrence, but it is worth noting that for future projects in similar locations, a larger contingency sum will be necessary.
- The final issue dealt with was that of drainage. At the outset, it was discovered that the records provided to us by the relevant water authority were inaccurate. We were informed that there was a possibility that the original drains ran directly under the proposed offices, creating further issues with regard to planning permission. This however, proved not to be the case and the drains ran through an adjacent builder's yard.

Reflective analysis of experience gained

This project has provided me with a great deal of experience, in an area that I would not otherwise have been involved with. Given the fact that I joined the council after the property located on the site had been demolished and the relevant funding secured, I was initially guided by the surveyor who I was to replace, and given a great deal of invaluable advice. After a few months, however, I was given full responsibility to obtain planning permission and take the project to practical completion.

The project was very interesting and provided me with great insight and enthusiasm for the development process. I have acquired a number of skills that are transferable to future development projects. I have also learned a number of lessons that will hopefully mean I will be able to avoid some of the issues that we encountered during the development of 23/27 Long Street.

One of these lessons is to keep a good relationship going with the planning department and try to involve them at every stage prior to obtaining planning permission. Delay at this stage can create great expense and in the case of this project it may have meant that we could have lost vital funding had the delay became an issue.

It is also to keep good lines of communication between all of the professional team in order that should any problems arise they may be identified and dealt with as quickly as possible avoiding delay. The appointment of a competent and efficient professional team is essential to any development project and as a result great time and care must be taken when considering these positions. A poor team can add time and cost to a project as opposed to avoiding it.

Finally, it is very important to have detailed surveys carried out before any details of a scheme are decided upon, as often these surveys will dictate the type of development possible for a particular site. Any surveys carried out prior to going on site will avoid delay and additional cost, or at least highlight the potential for additional work to be carried out prior to the contractor going on site.

The most important lesson which I feel that I have learned from this process, is preparation. The actual building stage of a scheme is a very small part of the project and 90 per cent of the work is carried out for a number of months prior to a contractor going on site. Once the contractor has gone on site, the majority of the work has been competed, and overseeing the development itself (as well as overseeing the relevant professional team) should be a relatively straightforward task. However, if a poor professional team has been appointed this task may become particularly onerous.

Appendices

The Appendices were: Appendix A – Photographs, Appendix B – Plan of the Town, Appendix C – Ordinance Survey Extract, Appendix D – Architect's Drawings and Plans, Appendix E – Architect's Brief, Appendix F – Local Plan Extracts, Appendix G – PPG 15 Extracts, Appendix H – The Council's Conservation Area Guide and Appendix I – Case Background. None of the Appendices are reproduced here.

Illustration of issues/possible interview questions

Q Can you explain the forms which had to be completed for the planning application, and the principal information needed?
Q Was there any retained land of the council, and why does this need to also be highlighted on the application/plans?
Q Can you please summarise the various aspects of national planning policy guidance relevant to the development – ie PPGs and anything else?
Q What were the planning conditions?
Q Were there any planning obligations?
Q You did not comment in any detail on the funding that was sourced, as it was not your responsibility, but as you have Economic Development as an optional competency, and are now working with a regional development agency, can you please explain the rules governing the award of grant funds?
Q How many car parking spaces were necessary in order to meet planning requirements/where is such a requirement prescribed?

Q Regarding the easement which could have been taken from the adjacent builder's yard, it was fortunate that you had established rights, but what would the valuation principles be if you needed to acquire the easement? Is the adjacent owner holding a ransom element? Are there are any alternatives?

Q What exactly is the criteria for the style of materials to be adopted?

Q If you considered the planners were being unreasonable in their request, what options would you have?

Q You did not have responsibility for preparing the initial development appraisal, but in delivering the development, must have regularly re-assessed the development appraisal. First, what were the key headings/content of the development appraisal?

Q What were the main sensitivities?

Q What was the contingency basis, and what does this reflect?

Q Did the contingency apply to all construction costs?

Q The council built the development itself, albeit with funding primarily from the regional development agency. What were the alternative options to the council building the development itself?

Q How would the developer view the risks, and how might this affect the development appraisal, and funding requirements if the site had been sold to a developer with a Development Agreement?

Chapter 15

Planning Appraisal

This chapter comprises a planning appraisal in respect of the alternative uses achievable for a university campus. The report is contributed by Guy Bransby.

Introduction

The subject of my critical analysis is the preparation of a Development Brief and subsequent planning strategy to facilitate the disposal of a university campus. This was part of a wider strategy to consolidate the university on its main campus, funded by disposal proceeds. This would allow the university to improve its educational offer.

I have selected this subject for my critical analysis as it highlights my specialist planning skills and demonstrates my expertise in the preparation of Development Briefs, providing town planning advice, consultancy skills, and housing strategy and provision. Furthermore, I was involved in the project from its inception to completion of the disposal, and will continue to be involved in the subsequent stages of the project, beyond my APC training period. Pursuant to this, I will be in a position to update the assessment panel on progress at my final interview.

This critical analysis is based on actual advice provided to the university, and, consequently, full regard has been given to confidentiality. Please refer to Appendix 1 for a copy of the client's letter of consent (not reproduced here).

Throughout the duration of the instruction to date, the lead partner in my firm (who is also my APC supervisor) checked all of my key correspondence and advice. Certification of the authenticity of this critical analysis by my APC supervisor and counsellor is attached at Appendix 2 (not reproduced here).

This report is structured as follows:

- A description of the site, setting out the university's objectives and my role in seeking to meet these objectives.
- Consideration of the initial planning considerations and detailed planning research required to formulate the Development Brief, which was a key part of the planning and disposal strategy.
- An analysis of my role in selecting the preferred developer, and formulating a joint planning strategy which met the university's objectives. This section also considers the way forward, as this is an ongoing instruction.
- Reflections regarding my role during the project, and what I have learnt in hindsight.
- Conclusions.

Site description

The university's campus is situated in a London Borough. Please refer to the site plan at Appendix 3 (not reproduced here).

The campus is 5.7 ha (14.1 acres) in size and is bounded by the River Thames to the east, and residential development on all other sides. The campus consists of a mix of education and ancillary buildings, including some student accommodation, which total approximately 9,000 m² (97,000 sq ft) in footprint, and over 12,000 m² (130,000 sq ft) in floorspace. Please refer to the site photographs at Appendix 4 (not reproduced here).

Client objectives

The university has a number of key objectives:

- To improve the university's educational offer and research capabilities to allow it to compete nationally with other Higher Education Institutions (HEIs).
- To increase participation in higher education by raising student and staff numbers at the university.
- To improve its links with the business and sporting community.
- To improve the quality of its buildings and the 'student experience' that is offered.

Please refer to the client description at Appendix 5 (not reproduced here).

To meet these objectives, the university needed to improve its current educational infrastructure. This required a fundamental reconsideration of the role of the university's estate and physical assets, as a major facilitator of educational provision.

Following advice from my firm, the University Council agreed to dispose of some smaller campuses, which included the subject site, to fund the consolidation and expansion of facilities on its principal campus. This would allow the university to operate more effectively and efficiently by maximising the use of finite resources.

Individual responsibilities

As a senior planner in the my firm's planning and regeneration department, I was required to provide advice on the following key aspects of the project:

- Undertake the background planning investigations.
- Consider the planning constraints and opportunities.
- Develop the planning strategy for the disposal.
- Prepare the Development Brief in conjunction with the local planning authority's officers.
- Provide our agency team with key planning information and assumptions to assist the development appraisals.

- Appraise the best offers and schemes submitted by interested parties in planning terms.
- Sit in on interviews with shortlisted developers to assist in the selection of the preferred developer.
- Develop a planning strategy with the preferred developer for the submission of a planning application for the redevelopment of the campus.
- Monitor the pre-application discussions between the local planning authority's officers and the developer.
- Ensure that the overage provisions within the sale contract were taken account of by the developer.
- Assist with the decant programme following the phased development.

Initial considerations

Following the university's approach to my firm, I established that there would be no conflict of interest. I was part of the team that made the presentation to the University Council in 2001 which secured the instruction, the terms of which were then agreed in writing. As part of this, the planning fee was agreed separately from the agency fee.

I then undertook a detailed site visit which involved:

- Becoming familiar with the surrounding area.
- Noting adjoining land uses.
- Taking photographs of key buildings and landscape features.
- Noting the access points.
- Establishing the key vistas.

Following this, I then undertook background planning research in the Borough's planning department. This included:

- Reviewing the Unitary Development Plan (UDP) to understand the policy designation for the site. Please refer to the UDP proposals map extract at Appendix 6 (not reproduced here).
- Considering the planning history to establish if a precedent had been set with regard to development and if there were any unimplemented consents or conditions.
- Perusing general UDP policy to advise the university of the thrust of key policy.
- Investigating the description of the listed building and the extent of the Conservation Area, as this would have a crucial bearing on the extent of demolition and redevelopment that would be allowed; and
- Having preliminary discussion with the local planning authority's development control and urban design officers to establish the preferred future use for the site. It was suggested that this should be predominantly residential with some community and education facilities and public access to the river and open space.

Consequently, I was able to advise the university that:

- There were buildings on the site with the potential to be 'spot-listed'. Clearly, any listing would affect the form of the redevelopment.
- There would need to be provision of large amounts of public and private open space within the scheme to maintain the green and open feel of the site.
- The 'lime avenue' of trees would need to be retained within any redevelopment.
- There were a number of Tree Preservation Orders (TPOs) that would need to be respected.
- Views from the river to the main listed building, Gordon House, would need to be retained and the landscape setting would need to be improved.
- The boundary wall would need to be retained around the site boundary as it contributes to the setting.
- There would need to be some public access to the site and the relationship of the site to the river should be improved.

A lot of these issues would later form the basis of the Development Brief.

Planning strategy

I advised the university that the preferred planning strategy should be to:

- Prepare a Development Brief for the site in conjunction with the local planning authority's officers and the local community to provide the purchaser with a better degree of certainty.
- Go to the market to dispose of the site subject to planning.
- Following the exchange of contracts, develop the planning strategy for the submission of planning applications to redevelop the site in conjunction with the preferred developer, with the university as joint applicants.
- Submit an outline planning application for the redevelopment of the entire site, with reserved matters applications to take account of the phased decant.

Detailed planning investigations

To allow the university to participate in the preparation of the Development Brief, and to provide sufficient information for my firm's agency department to appraise the value of the site, there were a number of planning investigations that had to be carried out. It was decided that the best approach would also require appropriate specialist sub-consultants to be appointed, and who would produce the following technical reports:

- A transport assessment to establish vehicle movements, congestion points, necessary transport infrastructure improvements, car parking arrangements and green travel initiatives.
- An arboricultural assessment to identify those trees that could be removed and those that would need to be retained.
- A listed building report to ascertain if there would be any merit in seeking to

have certain buildings delisted and to establish if any other buildings could be capable of listing in the future.
- An archaeological assessment to establish if there are any archaeological remains within the site, which would need to be protected or removed in advance of any redevelopment.

I was expected to manage these sub-consultants and their inputs within the required time scales, and with respect to the university's own guidelines.

The development appraisal

Following my background planning investigations and the completion of the sub-consultants' reports, the agency team was in a position to undertake a market appraisal of the site. This relied on assumptions including:

- The preferred and most sustainable mix of uses for the site.
- The net developable area.
- The amount of open space.
- The affordable housing requirement.
- Building density and height.
- Future use of the listed building.
- Possible section 106 contributions.
- Amount of landscaping required.
- Permitted car parking spaces.
- Access.

This allowed the agency team to provide the university with an indicative market value, which led to the decision to go to the market in early 2002. I was not required to undertake the development appraisal or have detailed input into the valuation work because of issues of client confidentiality and the need for me to focus on the Development Brief and the wider strategic planning aspects.

Preparation of the Development Brief

I felt that the best approach would be to engage with the local planning authority's officers at the earliest opportunity to prepare the Development Brief. This would allow the university to influence negotiations, and aim to ensure that the Development Brief was flexible enough for the preferred developer. At the same time, the local planning authority would be able to provide some certainty to members and the local community on the type, height and location of any future development.
   I was involved in a number of the key stages:

- Preliminary discussions with the local planning authority's officers to set the parameters of the Development Brief.
- Drafting key sections in conjunction with the local planning authority's officers in advance of the first draft.
- Submitting representations on the university's behalf on the first draft of the

Development Brief – objecting, amongst other things, to the height constraints and onerous affordable housing requirement.

- Redrafting certain sections of the brief to allow a second draft to be taken to Planning Committee in order to secure member approval to go out to public consultation.
- Engaging with members to set out the rationale of the brief, and supporting the university's key requirements following disposal of the campus.
- Organising and attending a public meeting in June 2002 to gauge the reaction of the community to the Development Brief. This was done with the support of the preferred developer.
- Further representations submitted on the university's behalf, countering the community's campaign to have part of the site retained in educational use (which would affect the university's receipts).
- Maintaining a close liaison with the preferred developer to ensure that its representations supported the thrust of those submitted by the university.

At the time that the preferred developer had been chosen and contracts had been exchanged, the Development Brief was still to go to public consultation. Consequently, the planning strategy needed to reflect this. As a result, a close working relationship developed between the preferred developer and myself.

The preferred developer

My firm's agency team put the site on the market, and I became aware that there was considerable interest in the campus. Please refer to the site particulars at Appendix 7 (not reproduced here). The agency team received over 150 initial inquiries, which were followed up by over 50 interested parties purchasing marketing packs which contained a lot of the information that I had collated.

Interested parties were then invited to submit best offers and a redevelopment scheme in order to allow the university to select its short list. I was asked to appraise these schemes, which I did by:

- Assessing each scheme in light of the Development Brief, which was now at final draft stage, and the UDP.
- Dismissing those developers whose schemes did not meet the criteria of the brief and/or UDP policy.
- Discounting those developers whose schemes, while meeting the objectives of the brief, were ill-conceived. This is a key consideration, as the university will maintain strong links in the borough after decant and would not want to support a scheme that would be unpopular locally.
- Draw up a final short list of developers who should be invited for interview on the basis of the schemes put forward and which was considered by the agency team in conjunction with the financial offer.

Following this, I was part of the panel, comprising the university and my firm, who interviewed the four short-listed developers. Having scrutinised the different redevelopment schemes, my role was to question the developers on their planning strategy to achieve success, understand the planning rationale underpinning the scheme design and layouts and come to a view as to whether the developer would be one that the university could successfully work with to achieve planning permission.

The preferred developer was selected following detailed consideration of a matrix that I had devised. This can be found at Appendix 8 (not reproduced here). Contracts for the sale of the campus were exchanged with the preferred developer in May 2002. My firm had, and still has, a key role to play.

Development of the planning application

Following the disposal of the site, the preferred developer was able to play a full part in the finalisation of the Development Brief. I advised the university and preferred developer to present a 'united front' when liaising with the local planning authority and the local community on the brief, which culminated in both parties hosting a public meeting in June 2002. This approach was well received and as anticipated, helped to minimise public objections to the Development Brief and the planned redevelopment of the site.

Taking this forward, borough members approved the Development Brief at its Overview and Scrutiny Committee in September 2002. This provided the development framework for the preferred developer's redevelopment masterplan. The approved Development Brief is at Appendix 9 (not reproduced here).

The preferred developer was seeking to maximise receipts for itself, but also for the university through an overage provision within the contract. This stipulated that should planning permission be obtained for a number of units above a certain threshold, the university would be due additional payments – in effect, a profit share arrangement. The developer was to use best endeavours to maximise density and the floorspace on the site, while at the same time respecting the site and surrounding area.

It was agreed that it was vital to have the support of the local planning authority's officers from an early stage and to maintain the close working relationship that had been fostered. A number of pre-application discussions were held with not only the borough. This was to try and achieve the support of the local planning authority's officers to ensure that the application would be taken to committee with a recommendation for approval.

The developer and the university were able to submit an application in December 2002 as joint applicants.

Way forward

In September 2002, my firm and the university agreed that I would be seconded to the university for one day a week to work in its Estates Department, assisting the Director of Estates and Pro-Vice-Chancellor for Development on all estates matters. With regard to this project, this will allow me to:

- Assist the university in monitoring disposal receipts to ensure that they are maximised and timely, in order to meet cash flow requirements.

- Accompany the preferred developer at all meetings with the planning officers on the university's behalf, in order to ensure that its interests and reputation are being protected.
- Monitor the preferred developer's preparation of its planning application to ensure that it provides for the overage provisions set out in the sale contract.
- Advise on the decant activity that will see the university vacate the campus in two phases, but ensuring that the campus still operates effectively to maintain staff and student recruitment and retention.
- Monitor the financial and tax implications pursuant to 'windfall' monies secured as part of the overage.

Please refer to Appendix 10 (reproduced at p000) for a chronology of events and the timing for the next steps.

Learning outcomes/reflective analysis

This project exposed me to the following key skills areas:

- The wide role that planning plays in the development process.
- The processes involved in preparing a Development Brief for adoption.
- The use of development appraisal as a numerical justification for the disposal strategy adopted.
- Application of consultancy skills to an agency/commercial environment.
- A greater understanding of the housing market.
- Familiarity with selecting and interviewing interested developers.
- Detail of sale contracts and overage provisions.
- The requirement for the university to achieve the best price reasonably obtainable in the current market.
- Secondment to an educational institution, which has allowed me to become familiar with the client side of property services and the workings of a large educational institution.

However, it is vital on any project to learn from experience and in hindsight, I would make the following comments:

- I could have made greater use of information technology for research purposes, such as the Focus and EGi databases.
- Development appraisal is very sensitive to assumptions and therefore, is meaningless without quality inputs.
- The housing market is sensitive to a number of local factors, as well as the state of the wider economy.
- It is important to understand the working relationship between the consultant, sub-consultants and clients, particularly with regard to communication lines and fee billing.
- I was unprepared for the complexities involved in the disposal process and preparation of sale contracts. At times, the level of information required by solicitors was daunting.

- As the valuation was for marketing purposes it was not a formal 'Red Book' valuation.
- I am now aware how crucial timing and programming is not only to securing best value, but to ensure successful construction and phasing of decantment in the latter stages of a project.
- At a number of stages throughout the project, it was important to make small concessions and compromises in order to allow progress to be maintained.

Overall, I have improved the following skills sets:

- Considering conflicts of interest and understanding quality assurance procedures.
- Confidence in handling client relationships.
- Team working and cross fertilisation of ideas and strategies.
- Verbal communication, presentation and negotiation skills.
- Detailed report writing.
- Technical ability in planning and development appraisal.
- Professional awareness.
- Responsibility to the client on behalf of my firm and its sub-consultants.
- Lateral thought and creativity required to become a strategic consultant.
- Time management, organisation and the ability to prioritise.

Appendices

The Appendices were: Appendix 1 – Client's Consent, Appendix 2 – Certification, Appendix 3 – Site Plan at 1:1250 Scale, Appendix 4 – Site Photographs, Appendix 5 – Client Description, Appendix 6 – Unitary Development Plan Proposals Map Extract, Appendix 7 – Site Particulars, Appendix 8 – Bid Matrix, Appendix 9 – Approved Development Brief, and Appendix 10 – Chronology of Events. Appendices 1–9 are not reproduced here. Appendix 10 is reproduced below.

Appendix 10 – Chronology of events

July 2001	University Council confirms intention to dispose of the campus.
August/September 2001	Incentivised agency fee arrangement agreed between my firm and the university.
September 2001	My firm undertakes soft market testing.
October 2001	First draft Development Brief agreed between the borough and my firm.
November 2001	The university's solicitors investigate the legal title.
December 2001	Planning investigations completed by my firm and its sub-consultants, including transport study, arboricultural assessment, listed building report, archaeological assessment, topographical survey, drainage survey, ecological survey, and geotechnical/environmental survey.

January 2002	Marketing pack completed by my firm's agency team.
February 2002	Final draft Development Brief agreed with the borough and put out to public consultation.
	Formal marketing of the site commences.
March 2002	My firm generate interest in the site, discuss it with potential bidders, undertake site inspections and seek 'best offers'.
May 2002	Offers received and evaluated.
	Prospective buyers interviewed.
	Contracts exchanged for the disposal of the entire campus with preferred developer.
June 2002	Public meeting on the Development Brief.
July 2002	My firm and preferred developer establish planning strategy for the redevelopment of the campus and engage with the local planning authority's officers.
September 2002	Development Brief approved by the borough's Overview and Scrutiny Committee.
January 2003	Target date for the preferred developer to submit the planning application for the redevelopment.
April 2003	Planning permission granted.
September 2003	Completion of the sale of the whole campus.
	The university will decant from the southern part of the campus as part of phase 1.
May 2005	The university will decant from the northern part of the campus as part of phase 2, thereby completely vacating the campus.

Illustration of issues/possible interview questions

Q What was your personal involvement in the project, and who did you report to?

Q What does your secondment at the University involve?

Q How would you advise a developer who was aware that its buildings may be listed?

Q What are the developer's prospects of success for redeveloping the campus?

Q What were the planner's views on retaining the listed buildings?

Q How might you achieve member approval for the redevelopment scheme?

Q How did the develop select its RSL?

Q When considering the matrix to select preferred developers, how was the balance struck between best price and a good scheme?

Q How would you deal with a footpath, were one to be found on site?

Q Will the developer's application be referred to the Secretary of State?

Q You spoke about the importance of a form of development being secured that would not be unpopular locally. How did you secure any such protection through the sale contract (preventing consents being sought for other schemes/variations)?

Q How were the overage provisions drafted? How do they work?

Q What weight does the Development Brief carry in terms of the consideration of the developer's planning application?

Q How did you embrace the concept of 'best consideration'/'best value' in view of this being a government client?

Q Can you please talk us through the merits of the offer and scheme that was selected, and also make comparison with key elements of any other offers?

Q Were you instructing the agency team, or were they making their own decisions. The reason I ask this is the volume of queries received. Would it have been preferable to circulate details to a number of selected developers who your agency team know had the track record and capacity to perform?

Q You mentioned that you did not undertake the development appraisal personally, but what were its key elements?

Q Where did the principal sensitivities lie?

Chapter 16

Compulsory Purchase and Business Valuation

This chapter covers compulsory purchase and the valuation of business interests. It is included in order to highlight the interrelationship between property interests and business valuation. The report is contributed by Rachael Jones.

Introduction

The subject I have chosen is a compulsory purchase compensation case where I am acting for the claimant. Compulsory purchase is one of my optional competencies, and one within which I have gained considerable experience during my training period.

This particular instruction demonstrates the knowledge and confidence I have gained within this subject. Consent to carry out my critical analysis on this instruction has been obtained from the client, however, I have agreed not to disclose the name of the property or the company.

Instruction

The property is a former railway goods yard in London owned by Railtrack. It consists of two levels, and when previously in use, trains came in on the upper level supported by two separate viaduct structures. Goods were moved between the upper and lower levels by a lift, and transferred to and from lorries at the lower level. Operational use ended in the mid 1960s.

The client acquired a lease from Railtrack in 1999 for the lower level of the goods yard measuring 2.175 ha (5.374 acres) and comprising mainly railway arches and access ways formed by the viaducts. The lease was for 10 years and excluded from the provisions of sections 24–28 of the Landlord and Tenant Act 1954. In accordance with standard Railtrack practice the landlord could terminate the lease at any time with six months' notice.

The client transformed the goods yard with his own investment, and with various grants, to provide flexible accommodation for a mix of uses, including workshops, studio space, leisure facilities and other commercial uses – and let the space to five tenants in 1999. Other arches remained vacant, as further work needed to be carried out before they could be occupied (and there were a number of parties interested in these).

Notice to Treat and Notice of Entry were served in November 2001 by London Underground Ltd (LUL) for the extension of one of their lines under the relevant order. This covered the whole of the goods yard.

My firm's instructions were to provide advice on both the acquisition procedure and the client's claim, and to agree compensation with the acquiring authority

(LUL). The fee was agreed in writing with the client and a conflict of interest check was carried out.

Key issues

The key issues/tasks for the instruction are highlighted below:

- Provide advice on the compulsory purchase procedure.
- Prepare the compensation claim, having regard to the six-month break clause within the lease.
- Negotiate and agree compensation with LUL.

Initial investigations

My first job was to establish the details of the client's lease and all sub-leases. I then carried out a property inspection and with the aid of floor plans measured the goods yard in accordance with the RICS Code of Measuring Practice (GIA of each lettable space). Finally, I contacted the relevant local authority to obtain the planning history for the site, establish whether any elements were listed or within a conservation area, and establish whether the site was in the authority's register of contaminated land.

These investigations were necessary to provide an initial feel for the likely issues, options and opportunities available. I then considered the client's options, both as to the acquisition and the subsequent compensation claim.

Acquisition

The client's lease allowed the landlord to terminate at any time on six months' notice. It was understood that Railtrack would develop the site in the future, however, the client and I believed that the chances of Railtrack actually terminating the lease in the near term were minimal – indeed the client would not have invested if he thought otherwise. Nevertheless, it was obvious that LUL would argue for an assumed shorter period – ie an early termination of the lease. The client therefore ran a very real risk of getting minimal compensation. I advised the client to delay giving up possession so that he could maintain his income stream for as long as possible.

However, the client had, before we were instructed, already worked on the assumption that LUL would take possession on the date in the Notice of Entry (three months from the date of the Notice (ie February 2002) and notified all service suppliers of this.

As the option of remaining in occupation was not available, I informed LUL that we were basing our claim on the assumption that they would take possession in February 2002.

A number of the client's sub-tenants were still in occupation, and wished to stay as long as possible. LUL requested that after the date of possession, the client stay on to manage the site for an agreed fee in order to look after the remaining sub-tenants. My client agreed to this, and I negotiated a reasonable fee. However, LUL backtracked on their decision and proposed an immediate assignment of the client's lease. I instructed our solicitors to organise this as quickly as possible. Unfortunately, there was some delay by LUL and the assignment was not agreed until 1 July 2002 (the revised date of possession).

Claim

I then set about considering the claim. The standard approach to the assessment of compensation would have been to assess the open market value of the leasehold interest plus any disturbance.

However, this assessment of compensation would not have reflected the actual loss suffered by the client, as there may be a limited market for a lease with a six-month break clause, and the value/compensation would therefore be minimal.

In order for the client to receive appropriate compensation, my line manager advised that the compensation claim should be based on the value of the client's interest as a business plus any disturbance costs, i.e. total extinguishment of the business. This enabled me to value on a 'value to owner' basis rather than a 'market value' basis.

Taking this into consideration, I decided to value the business with the use of a discounted cash flow. This method enables many relevant items to be included in the valuation in a format which allows ease of understanding and a number of variations to be included. There are various sources of income for this business, and various increases in the income per annum, therefore, this seemed the most appropriate method.

I based the discounted cash flow on the assumption that in the 'no-scheme world', the client would be able to remain in occupation until the end of the lease (ie 2009). As mentioned above, the client and I believed that the chances of Railtrack actually terminating the lease to develop the site in the near term were minimal. As the goods yard was net income producing for Railtrack, they would not wish to risk losing this income. There is also evidence by way of the various leases granted in the railway industry that despite such break provisions, the landlord very rarely determines the lease, and tenants are prepared to accept a break provision and still invest in the property. (Tenants may, however, seek to specify precise reasons for the exercise of the break provision, such as the property's necessity for operational railway use.)

I did not consider that the lease would be renewed in 2009, as it was going to be difficult enough to argue for the full term. Additionally, the lease is excluded from sections 24–28 of the Landlord and Tenant Act 1954.

Calculation of the claim

In order to compile the claim, I had constant contact with the client by letter, email and through telephone conversations. A copy of the claim is shown at Appendix A (reproduced at p160) and includes the following heads:

Value of business: Loss of income.
Disturbance costs: Pre-possession losses.

Directors' and staff time in dealing with the claim.
Professional fees.
Grant clawbacks.

Loss of income

This head of claim includes the discounted cash flow, calculating the value of the business – shown at Appendix B (not reproduced here). The NPV of net income is calculated for the period from the date of possession to the end of the lease (1/7/02–31/3/09). In order to do this I first calculated the total gross income per annum as shown in the table at Appendix C (not reproduced here).

Gross income

I included the income from the let space which I established from each of the sub-tenant's leases increases by 10% on the prior years' rent. I took into account when the rent increases would take place throughout each year, depending on the rent commencement date, resulting in rents 'a' and 'b' as shown in the table at Appendix C.

I also included the potential income from lettable space, as the client informed me that there were 178 parties interested in taking space at the goods yard. Arches 9, 11–18 and 24–30 would have been ready in November 2001, February 2002 and on 7 January 2002 respectively. I contacted the client to establish appropriate rents for this space, which I checked against existing lettings. The northern arches (24–30) would have let for £5.38 per m^2 in 1999, the southern arches (9 and 11–18) would have let for £32.29 per m^2 in 1999 – all with rental increases of 10% per annum in accordance with the other lettings. Again, I calculated the rent per month taking into account when the rent increases would take place throughout each year as shown in the table at Appendix C.

Finally, I included income from filming and events held at the goods yard, shown in detail at Appendix D (not reproduced here).

Net income

The net income was calculated by deducting the costs to the client, which included rent payable to Railtrack (as shown in the table at Appendix E – not reproduced here).

A further cost is the commission payable to Amazing Space who acted as an agent for the client, and organised filming and events on site. This is shown in detail at Appendix D.

I also included the cost of making ready the lettable arches. No works needed to be carried out to arch 9, works needed to be carried out to arches 11–18, and tenants were to carry out their own fit out work to arches 24–30 which was reflected in the lower rent payable for the northern arches. I requested costings from the client for arches 11–18 (as shown in the discounted cash flow at Appendix B). The client confirmed that the grants would not have been used for this work.

I did consider including the legal costs for drafting and agreeing the leases for the lettable space, and for the court orders for excluding the leases from the provisions of sections 24–28 of the Landlord and Tenant Act 1954 but decided against it. The client's solicitors confirmed that the client would have drafted the leases, and the remaining costs were recovered from the in-going tenants.

I considered including agents' fees, however, the client confirmed that they dealt with lettings themselves.

Finally, I contacted the client to establish the operating costs at the goods yard – for example management costs, electricity, gas. The client informed me that all operating costs were covered by the service charge payable by each of the subtenants. Therefore I did not include these costs.

NPV of net income

After deducting the appropriate costs, I was able to calculate the total NPV of net income shown in the table at Appendix B. I decided on a discount rate of 7%, bearing in mind my assumption that the business would have continued for the full term of the lease and that the client currently pays interest on overdrawn bank balances at 2% over base rate (currently 4%) – namely 6%, and the rate of interest paid by the client on director's loans is 7%.

Pre-possession losses

This is divided between loss of income and costs incurred. Arches 9, 11–18 and 24–30 could have all been let before the date of possession. Therefore I included the loss of income from these rents in the claim at Appendix A.

One of the five original tenants ceased to pay their rent from the end of November due to the compulsory acquisition. I therefore included the rent from December to the date of possession as a loss of income at Appendix A.

The only cost incurred pre-possession was for the termination of a contract for a CHP unit installed at the goods yard. The client received an invoice for the removal of the unit and the cancellation of the contract for the amount shown in the claim at Appendix A.

Directors and staff time in dealing with the claim

At the beginning of the instruction, I wrote to the client to request that each person within the company dealing with the CPO keep a note of their time spent and to include a description of the work carried out. The time spent is shown in the claim at Appendix A.

Professional fees

The professional fees included legal fees, surveyors' fees and other professional fees as shown at Appendix A.

Grant clawbacks

As mentioned above, the client received a number of grants in order to redevelop the goods yard (and which were for work carried out to the southern arches). I obtained copies of all of the grant agreements from the client and summarised each of them. Two grant agreements contained clawback clauses which required the grants to be repaid on a pro rata basis, and the client informed me that they made similar agreements with the remaining grant bodies. I calculated the possible clawback amount as shown in claim at Appendix A. Currently, the grant bodies have not requested repayment of the grants, nevertheless, the liability remains with the client.

Current position

I requested an advance payment of compensation as provided by section 52 of the Land Compensation Act 1973. The advance payment is based on 90% of the acquiring authority's estimate of compensation, and LUL offered £40,000. From this payment. I established that LUL's compensation estimate was considerably lower than ours, I understood that LUL assumed that Railtrack would have developed the site immediately in the no-scheme world. Nevertheless, there were a number of heads of claim which should have been included within their estimate – for example the grants and the cost for the removal of the CHP unit. I contacted LUL to highlight these points, they considered what I said and then made an additional payment of £146,124.

After a number of discussions with LUL regarding the claim, the client, solicitors, my line manager and myself decided that we were not going to agree compensation and made a reference to the Lands Tribunal.

A pre-trial review was held in November 2002 where it was decided that we submit a Statement of Case and for LUL to respond to this statement. We submitted our Statement of Case in December 2002 and are currently awaiting a response from LUL.

Critical appraisal

The client has been happy with our progress so far, however, the work in one area of the instruction which could have been improved was the contact and relationship we had with LUL. We had various meetings with the surveyor at LUL where we explained the basis of our claim. However, we received no information from LUL on how they were approaching their assessment of the compensation. Once we referred the claim to the Lands Tribunal, LUL changed the surveyor dealing with the claim. This has hindered the communication further, as what little relationship had built up has now been lost. So far we have failed to get any response from the new surveyor/witness. Unless this changes in the near future, it will be very difficult to comply with the likely Lands Tribunal directions concerning expert witnesses agreeing facts and other matters.

Another area open to improvement was the acquisition process. As mentioned earlier, I would have advised the client to remain in occupation for as long as possible so that the company could maintain their income stream. However, the client assumed, before we were instructed, that LUL would take possession on the date in the Notice of Entry – therefore any benefits which could have been gained from this were lost.

The basis of the claim (as a business) may cause problems at the Lands Tribunal as the standard approach to assessing the compensation would have been the open market value of the leasehold interest plus disturbance costs. However, if I had valued on this basis, it would not have reflected the actual loss suffered by the client as there may be a limited market for a lease with a six-month break clause. My line manager believed this was not a normal property investment. The client was running the goods yard as a business, as he received income from the occupiers and had a close relationship with each of them. The client also arranged filming and events to be held at the goods yard. Therefore, we decided to assess the compensation as a business plus disturbance costs, thus reflecting the actual loss suffered by the client. This will be an issue at the Lands Tribunal which my line manager acting as expert witness will have to argue.

If the Lands Tribunal does not agree with our view, then we could argue that there would be some value on the open market. The client, grant bodies and subtenants invested considerably in the goods yard, all assuming that they would be in occupation for the whole term. If these parties were all prepared to take on the risk of the six-month break clause, then another notional person may also have been prepared to take over the lease, therefore creating a market.

Another point which will be argued at the Lands Tribunal is 'if and when Railtrack would have terminated the lease in the no-scheme world to develop the site'. A colleague within the planning department is researching this. Initial conclusions are that in order for Railtrack to have obtained planning permission for the site, the local authority would have required a rail link to support the development. In the no-scheme world, there would be no extension of the London Underground line. A new station could have been provided on the Central Line, however, committee reports indicate that the Central line is overcrowded, and would need some relief before a new station could be provided. Crossrail would provide relief, however, this has yet to obtain a Transport and Works Order. These factors would delay any development on the site in the no-scheme world, enabling the client to stay in occupation. Nevertheless, if we assume that Railtrack applied for planning permission now, there is always a period of time before permission is granted and development commences on site, requiring tenants' leases to be terminated. My line manager envisages that it would take approximately five years, therefore the client would be able to remain in occupation for the majority of the lease.

Reflective analysis

Although I had poor communication with the acquiring authority, I believe I have improved my communication skills through the contact I had with the client. I have consulted with the managing director, account manager and the site manager of the company since the beginning of the instruction, at meetings, through telephone conversations and email. After having such close contact with the client, I now feel more confident when dealing with other clients.

I have improved my skills as a valuer, as this is the first compensation claim I have prepared with very little guidance and supervision. My line manager was available whenever I needed assistance, however, I carried out the work myself. I have become more competent with Microsoft Excel after compiling the compensation claim on this system.

I have assisted on compensation cases which have been referred to the Lands Tribunal, however, this is the first case where I have prepared the claim myself and it has then been referred. Since the case has been referred, my calculations have been subject to close scrutiny from the barrister, the solicitors, the client and my line manager who is acting as expert witness. My line manager has to rely on the figures I have produced, and understand all the decisions and assumptions I have made. Through this process, I have become more confident of my own decisions and abilities, and learnt the importance of providing evidence to support the claim.

I have really enjoyed working on this instruction and it has been exceptionally good experience being heavily involved in a case which has been referred to the Lands Tribunal. I have benefited from the experience gained through this process and I look forward to hearing the response to our Statement of Case from LUL.

Appendices

The Appendices were: Appendix A – Compensation Claim, Appendix B – Table Calculating the NPV of Net Income, Appendix C – Table Calculating the Total Income Per Annum, Appendix D – Filming/Events Income and Amazing Space Commission, Appendix E – Rent Payable to Railtrack. Appendix A is reproduced below. The other Appendices are not reproduced here.

Appendix A – Compensation Claim

Loss of income	
Net after allowing for necessary costs	£1,228,140
Pre-possession losses	£174,950
Director's and staff time	£24,820
Professional fees	£53,234
Grant clawbacks	£544,712
Total Claim excluding VAT and interest	£2,025,855
VAT	£354,525
Total claim including VAT but excluding interest	£2,380,380
Interest at statutory rate (to be calculated)	

Illustration of issues/possible interview questions

The detailed business valuations are excluded from the appendices covered as this chapter is included in order to provide an insight.

Q You referred to measurement on a GIA basis. How did you take account of the arch structure/was an element of the arches below 1.5m in height owing to the curvature?

Q For a railway goods yard adjoining an operational line, what aspects may limit use, development etc of the site compared with non-railway sites?

Q What is the actual impact and affect on the tenant's rights if the lease is contracted out?

Q If Railtrack could terminate the lease at six months' notice, what is the reason why your client can still make a claim based on a more substantial security of tenure?

Q If London Undergound had acquired Railtrack's interest, could they have terminated the lease with your client and avoided having to pay compensation?

Q Regarding total extinguishment, was it necessary for you to look for alternative premises?

Q What process did the acquiring authority undertake in order to implement the CPO?

Q Did you check that it was valid/what might your colleagues have been doing to protect the client's/claimant's interest?

Q Can we run through the six valuations rules for compulsory purchase, and see how they might apply to the property, or if not, how they might apply in other situations?

Chapter 17

Compulsory Purchase and Planning Issues

This chapter considers compulsory purchase and planning issues. The report is contributed by Kate Terriere.

Introduction

The subject of this critical analysis is a report on the planning assumptions relevant to a claim for compensation for the compulsory purchase of a service station on the A1 at Alconbury, Cambridgeshire. This project has been selected as it is representative of my experience in this area as well as overlapping with other areas of work in which I have been involved.

This report will outline the project, highlighting the key issues and objectives. The various options relevant to the solution of the project will then be discussed as well as a justification of the selection of the preferred option. The report will conclude with a critical appraisal of the outcome of the project and a reflective analysis of my experience. Commercially sensitive information has been omitted.

The project

This report is based on the compulsory acquisition of a plot of land on the A1 by the Highways Agency for the construction of new trunk roads at Alconbury. Powers were granted under 'The A1 Trunk Road (Alconbury to Fletton Parkway Improvement) Compulsory Purchase Order No 19 (1994)'. The subject land was a northbound site fronting the A1, north of Alconbury near Huntingdon, Cambridgeshire (see Appendix 1 – not reproduced here).

The client instructed my firm to produce the claim for compensation for the compulsory acquisition of their land. Our valuation department were dealing with the claim in conjunction with our automotive and roadside department. My role in the project was to assist in producing the 'Expert's Report on Planning Assumptions' associated with the claim. As the project progressed, it was clear that the success of the claim would be planning driven, and this therefore placed significant emphasis on the planning assumptions report.

Advance payments were made, but the amount of compensation was disputed. The case was to go to the Lands Tribunal on Monday 19 November 2001 where the whole claim would be assessed.

The date of publication of the Compulsory Purchase Order was 22 October 1991. The valuation date was agreed as 5 December 1995.

The site

The site totalled 1.33 ha (3.29 acres). The site owners also owned a site adjacent to the southbound carriageway of the A1 just north of the subject site. The northbound site was let to Wodehams Garages in 1956 and remained in their occupation until 1990 when the leasehold interest was purchased in order that both the northbound and southbound sites could be redeveloped.

The site had a planning history of permissions for 'transport services'. In July 1989, the owners submitted a planning application for a petrol-filling station, lorry/coach park, car parking for 85 cars, a 40-bed Travelodge, and Little Chef and coffee shop. The application was revised in May 1990 further to discussions with the local planning authority (Huntingdon District Council). The amended application was recommended for approval, but permission was refused at the direction of the Secretary of State for the Department of Environment and Transport. This will be discussed in more detail later in the report.

Key issues and objectives

When putting together a claim for compensation, there are a number of factors to be considered. They mainly revolve around valuation and planning. It was clear from the beginning of this case that the other party's expert statements would differ significantly from our own.

I was involved in the planning assumptions, but not directly in any valuation. However, it was important that I understood the issues regarding valuation so these will be discussed briefly. The key issues relevant to each area are:

Valuation: Method of valuation to be adopted.
                   Estimating future sustainable fuel throughput.
                   The standard and form of petrol filling station and facilities.
                   The 'turn-in' ratio.
Planning: What assumptions should be adopted?
                   The no-scheme world.
                   What evidence is available?
                   What assumption maximises our claim?
The claim: Improving the claim without compromising our position – the strength of our case.
                   Negotiating with the opposition.
                   Supporting each option with a planning case.
                   Which option to run with?

Objectives

Our main objective was to claim the correct amount of compensation for the client, and be able to justify this amount through the method of valuation and the planning assumptions adopted. The amount had to be realistic and capable of being scrutinised by the Lands Tribunal. It was important that our claim did not compromise our client's position, and risk losing the case and potentially a substantial amount of money.

The Expert Witness has a duty to the court to assess all the evidence to produce a claim that is fair. However, they are entitled, where there is doubt, to look at the top end of the scale and therefore improve the amount claimed on behalf of the client.

Valuation

While I had a very limited role in the valuation aspect of drawing up the claim, I attended meetings and conferences with counsel where valuation issues were discussed. It was important that I had an understanding of the different values, as the planning assumptions would be used to justify the amount claimed.

From the beginning of the case, four different bases of valuation were identified, and it was decided with the other party that it was these four valuations that would try to be agreed. They were:

A   Petrol filling station (PFS), lodge and restaurant.
B   PFS, restaurant plus expansion land.
C   Redevelopment of modern PFS and café, plus expansion land.
D   Replacement of existing PFS and café, plus expansion land.

The valuations were based on the type of facility assumed at the service station, and therefore the estimated fuel throughput. The different valuation scenarios produced different amounts, of which Option A generated the highest potential claim.

Petrol filling stations tend to be valued by assessing the future sustainable fuel throughput by analysing the existing level of fuel sales in relation to the operating conditions on site. Potential fuel sales are assessed with reference to traffic flow and 'turn-in ratio'. The ratio will differ considerably between an urban filling station and a motorway service area, but generally ranges between 2% and 5%. The annual throughput can then be ascertained by adopting an average purchase of gallons.

The main issues that were contended by each party related to throughput – the number of gallons and price per gallon to be assumed under each valuation. The type of facility assumed would affect the turn-in ratio, the value of the site and therefore the potential claim. Other issues related to 'expansion land' and 'hope value' – whether to assume any and if so, what value could be attributed to them.

Planning assumptions

When assessing compensation, the relevant interest can be valued with the benefit of a planning permission. Planning assumptions are dealt with under sections 14 to 17 of the Land Compensation Act 1961 (LCA), as amended by the Planning and Compensation Act 1991. Each section sets out the different assumptions that can be adopted when assessing compensation. A further explanation of these can be found in Appendix 2.

The planning case

In producing the report on planning assumptions, I undertook a significant amount of research. It was necessary to research the planning history of the site as well as looking at the relevant Development Plans. I also researched Government guidance and circulars. Planning history records were copied and analysed. It was important to look not just at the various decisions but the reasons behind them. Details of correspondence could be used as evidence. The history of other service stations in the area was also researched.

Initially we were uncertain as to which planning assumption to adopt, and therefore my research needed to cover a period of time, concentrating on the valuation date and date of Notice to Treat. At all times, it was necessary to consider the evidence adopting the no-scheme world. In this case, the question to be answered was 'but for the widening of the A1 Trunk Road, what would have happened?' It was difficult to ascertain what may have happened at these different dates. For example, the relevant Development Plans changed over time as they were reviewed plus many of the policies referred specifically to the A1 works within 'the scheme'.

Following the investigation of the site's planning history and relevant Development Plans, I then helped draft the report on planning assumptions. This was basically an assessment of all the relevant information to establish what development, if any, would have been granted planning permission in the no-scheme world.

A copy of the final report on planning assumptions can be found at Appendix 3 (not reproduced here).

The options

The various valuation options have been highlighted previously. In determining what might reasonably have been expected to be granted planning permission, our planning case would be vital to the selection of which option to run with and present within the claim.

The most significant piece of planning evidence in our case was the planning application submitted by the landowners in July 1989. The application consisted of a petrol filling station, lorry/coach park, car parking for 85 cars, a 40-bed Travelodge, and Little Chef and coffee shop. This application was amended in May 1990 to 'redevelopment of existing petrol filling station and reconstruction of existing derelict café to form modern motorist facility'. The site was set back 20m, the Travelodge removed, and the car parking amended. However, the application was still refused in July 1990 for three reasons:

- The proposed development is premature to the completion of statutory procedures for the widening of the A1 Trunk Road.
- The existing facility at this site is no longer used and the site is in a near derelict state. Consequently, the proposed development would result in a significant increase in the number of vehicles using the existing accesses from the site to the trunk road. The additional slowing down and turning movements involved would be to the detriment of the free flow and safety of traffic on the trunk road. This is notwithstanding the applicant's intention to improve the existing accesses to current standards.
- As there are existing service facilities on the same side of the trunk road some four kilometres north and seven kilometres south of the site, there is no need for additional facilities at this location which would justify overriding the objection given in (2) above.

Although this refusal is significant, in the no-scheme world we must ignore any reference to the widening of the A1. The first reason given for refusal was that the proposed development was premature to the completion of statutory procedures for the widening of the A1 Trunk Road – this must clearly be ignored. The second reason related to traffic flow and safety on the trunk road. The amendments to the planning application had taken into account some of these issues, and in the no-scheme world these could have been resolved through further discussion and minor alterations. The third reason for refusal related to the need for the facility. It was considered that there were enough facilities already in the area. However, traffic growth was anticipated, but not as a product of any improvement works, so in the no-scheme world, this growth would have generated demand for service station facilities.

Little Chef was already operating from a site to the north and would relocate to the subject site to consolidate facilities. The planning and highways authorities were concerned about highway safety and the number of turn-in points along the A1. In the planning process, this issue would have been a negotiating tool – if the planning authority granted permission, they would reduce the number of roadside services and therefore the number of accesses from four (two per site) to only two. Although this issue would be relevant for any other planning application, it is difficult to replicate and prove in the no-scheme world.

We concluded that planning permission would have been granted for the amended application submitted by the landowners, and also for the original application. However, our view was not mirrored by the other party, who argued that there would have been no reasonable prospect of planning permission for a service area. They did concede that rebuilding of a filling station and café (those facilities established on the site since 1956) may have received consent.

The original planning application fits valuation Option A, and the amended application fits Option C. Although we had concluded that Option A would have secured planning permission, we had to consider whether we had enough evidence to present a strong case. Option A provided us with the highest amount of compensation, whereas the other party were stating that only Option D, ie replacement facilities, was viable. It was clear there were significant areas of disagreement between the parties, not just on the amount of compensation, but also on key principles such as 'turn-in' ratios, the need for sliproads as part of the development, and expansion land.

As part of the Lands Tribunal rules, each party has an obligation to agree as many facts and principles between them as possible, before reaching the hearing. The Tribunal can be expensive and the outcome is not ever guaranteed. Although, we were comfortable with our conclusions, we had to assess whether we were confident enough that our case would win at Tribunal. Following several conferences with Counsel, it was decided that it was too risky to claim Option A, and that our claim should follow Option B or Option C as a minimum.

Outcome of case

In the build up to the Tribunal hearing we were in negotiation with the other party to settle. As such, it was important that we did not select one option above any others and reveal this to the other party. We had to maintain our position and portray to the other party that we were confident that we could prove our case in claiming the optimum compensation for our client. Negotiations were held mainly through each party's solicitors while each expert witness (ie planning and valuation) dealt with their respective opposites. An offer was made by the Highways Agency in September 2001, but this was below our minimum position.

As the case progressed it looked likely that we would settle rather than proceed to Tribunal. There were tight deadlines to get all evidence catalogued, statements of agreed facts produced and expert reports produced. All our evidence still had to be prepared just in case we could not settle at an appropriate amount and we would be going to the Lands Tribunal.

The planning case became the driver in the amount of compensation we should claim. Both parties differed on their conclusions – we thought that planning permission would have been granted, and the other party thought not. The differences in planning assumptions produced quite significant differences in the amount of compensation claimed. As our case was not clear-cut we could not risk losing at Tribunal, and ending up with a minimal payment.

We had made an offer of compensation to the other party and were awaiting a response. The other party was arguing that no planning permission would have been granted, and development would not have taken place – therefore any offer they made would be less than ours. We pointed out that under section 15 of the Land Compensation Act 1961, Schedule 3 rights are to be assumed. This meant redevelopment of the existing facilities to a modern equivalent – reflective of our Option C. The other party conceded on this point.

The Highways Agency presented an offer to us for an amount greater than our valuation for Option C. We accepted, and the case was settled on the Friday (16 November) before we were due at the Lands Tribunal on the Monday (19 November).

Critical appraisal and reflective analysis

The case was relatively successful in that the compensation received was greater than that which we considered satisfactory. There was the risk, had we gone to the Lands Tribunal, that we would have lost the case, and ended up with less than was on offer (as well as having to pay the professional fees). Although the compensation received was not our optimum amount, the compensation was satisfactory to our client.

There were difficulties in negotiating with the other party throughout the case. Often it took a long time for the other party to respond to our correspondence or to questions or issues raised. Had it been possible to agree certain matters with the other party, our valuations may not have been so different. However, it is generally difficult to agree many issues in such circumstances, as the parties' respective objectives varied.

I played a significant role in producing the report on planning assumptions associated with the claim for compensation. I undertook the planning research, looking at planning history records, relevant government guidance and policy and various development plans. I did encounter problems with seeking information from the planning authority. Records were not easily accessed, and in order to obtain copies of relevant documentation each item had to be listed and separately requested. There were also delays in receiving these copies. In retrospect, I could have been more time efficient in the research I carried out.

With all the research I carried out, it was necessary to distil this information, and report the main facts and important issues. There was a wealth of information to go through, and at times it was difficult to filter out the important issues. However, it was also important that my research was thorough. At all times I also had to consider this information in relation to the no-scheme world.

I found it difficult to understand the concept of the no-scheme world when all the evidence clearly related to different aspects of the scheme. For example, planning policies would specifically relate to the widening of the A1, yet in the no-scheme world we have to ignore this. It was difficult to ascertain what might have happened had it not been for the scheme, when all the evidence had reference to the scheme or other events had been a consequence of the scheme.

I have learnt a significant amount from my experience in this case. My knowledge of the compulsory purchase compensation process has increased, and I now have an understanding of the different components of a compensation claim as well as the different roles within the team. I have also learnt the importance of concise and clear reporting of facts, and how their interpretation is crucial.

I have also learnt about the planning system. In trying to establish what may have received planning consent, it was apparent that there are differences between what policies say and what planning authorities actually do. In our case, policies were generally against development but the planning authority were in favour – it was only at the direction of the Department of Environment and Transport that planning permission was refused.

I played an important role within the team, having responsibilities from an early stage, and this continued through the negotiations which ran over a four-month period. I liased with colleagues within as well as the solicitor on the case. I learnt about negotiation and the significance of time pressures.

I worked directly with a partner, who was the expert witness, reporting my findings on a regular basis and discussing the various issues that arose. I was able to review my analysis of the information, discuss the different assumptions and assess which should be adopted.

My comprehension of the compulsory purchase process has increased. My work could have been more efficient, but I have now learnt the importance of interpreting information and the presentation of the relevant facts. I understand the concept of the 'no-scheme' world, and have also learnt that it is important not to get bogged down in all the information, and to be able to understand what is necessary to support our case.

I now have more confidence in my abilities and have been able to practice what I have learnt in assisting on another compulsory purchase case.

Appendices

The Appendices were: Appendix 1 – Site Plan, Appendix 2 – Explanation of Planning Assumptions under Sections 14 to 17 of the Land Compensation Act

1961, and Appendix 3 – Copy of Report on Planning Assumptions. Appendices 1 and 3 are not reproduced here. Appendix 2 is reproduced below.

Appendix 2 – Explanation of Planning Assumptions Under Sections 14 to 17 of the Land Compensation Act 1961

Section 14 of the Land Compensation Act 1961 states that any planning permission in force (outline or full) can be assumed at the valuation date. Refusal of other consents is not necessarily excluded as it is more a question of what would be reasonable to assume. Under subsection 5, it states that 'where the land is to be acquired for a highway it is to be assumed, for the purpose of determination under section 16 or section 17 of the 1961 Act, that no highway will be built on any other land unless it is already held or authorised to be acquired for highway purposes'.

Section 15 states that planning permission may be assumed for the development proposed by the authority. It also states that Schedule 3 rights (as amended under the Planning and Compensation Act 1991) are to be assumed. Schedule 6 of the Planning and Compensation Act 1991 states that it shall be assumed that planning permission would be granted for any development specified in paragraphs 1 and 2 of Schedule 3 of the Town and Country Planning Act 1990. Paragraph 1 (b) states the 'rebuilding, as often as occasion may require, of any building erected after 1 July 1948 which was in existence at a material date'.

Section 16 relates to development in an area of comprehensive development. For any land outside an area of comprehensive development, it may be assumed that permission for any land use specified in the Development Plan, or in accordance with the zoning of the land would have been granted. Within an area of comprehensive redevelopment, permission may be assumed to be available for any form of development within the range of uses proposed for the area. These planning assumptions are taken at the date of Notice to Treat.

Section 17 provides for an application to be made to the planning authority for a certificate stating what development (if any) might reasonably have been permitted if the land were not to be compulsorily acquired. This enables land that is not allocated to its most valuable potential to be valued with the benefit of a notional planning consent, if the section 17 application is successful. As a result of case law, a section 17 certificate is considered at the date when the Order giving rise to the acquisition was first published.

Illustration of issues/possible interview questions

Q Can we just clarify what the elements of the compensation claim were: market value of the land taken is clearly one, but what are the others?

Q You did not mention betterment, but how might this be relevant to the circumstances you described?

Q What is the position regarding surveyors' costs – how are they calculated?

Q What was your fee basis, and what where the key terms in the letter of instruction?

Q What are the rules on advanced payments, and are there any particular tactics/strategy than can be applied to secure advanced payments as soon as possible?

Q What determined the valuation date?

Q How does this affect planning assumptions, and having regard to your wider experience of planning, can you think of any situations where time scale could have a significant difference (also, how is Notice to Treat relevant here)?

Q What is the scope for 'hope value' to be reflected in a claim, and how can 'hope value' be quantified?

Q What about 'marriage value', and the scope to bring in neighbouring property to a development/valuation?

Q Regarding your point about expert witness having a duty to a court/tribunal, what is your experience in practice of how the respective parties might look to stress the more favourable aspects of their case?

Q You commented on the valuation approach to petrol filling stations. How has the market for petrol filling stations performed over recent years (noting the number of stations closing), and how does the valuation account for the potential for shop sales?

Q Regarding the no-scheme world, what legislation and case law covers this?

Q How significant is a 'widening' of a dual carriageway in creating a difference between the scheme world and no-scheme world.

Q What other elements of highways, infrastructure, etc, could have a significant effect regarding the scheme world and no scheme world?

Q One of the points raised regarding planning was that there were other facilities already in the area. Is this a valid planning reason, or simply a point as to commercial viability (ie what is the extent to which uncertainty over commercial viability affects planning)?

Q You mentioned that it was decided to not claim Option A at Tribunal because it was too risky, and follow Option B or Option C as a minimum, but would it be possible to claim A, but adopting a fallback position of B or C (and is the position influenced by having to work in the expert witness role)?

Q Does the Lands Tribunal have the ability to award costs against one of the parties?

Index

Printed and bound by CPI Group (UK) Ltd, Croydon, CR0 4YY

01/11/2024

01782614-0018